Thermal Energy Storage

Hafiz Muhammad Ali · Furqan Jamil ·
Hamza Babar

Thermal Energy Storage

Storage Techniques, Advanced Materials, Thermophysical Properties and Applications

Springer

Hafiz Muhammad Ali
Mechanical Engineering Department
King Fahd University of Petroleum
and Minerals
Dhahran, Saudi Arabia

Furqan Jamil
Mechanical Engineering Department
University of Engineering and Technology
Taxila, Pakistan

Hamza Babar
Mechanical Engineering Department
COMSATS University Islamabad
Sahiwal, Pakistan

ISBN 978-981-16-1130-8 ISBN 978-981-16-1131-5 (eBook)
https://doi.org/10.1007/978-981-16-1131-5

This Springer imprint is published by the registered company Springer Nature Singapore Pte Ltd.
The registered company address is: 152 Beach Road, #21-01/04 Gateway East, Singapore 189721, Singapore

Preface

Thermal energy storage becomes increasingly vital as renewable energy sources are currently catching up and speedy diminishing of fossil fuels. It is one of the major dawns in the field of thermal energy management. The usage of thermal energy in the form of sensible or latent heat is an effective method to store and utilization of energy. Thermal energy based on phase change materials can alleviate the energy crisis due to high energy output and high energy density. So, the emphasis of this book is to cover various outlooks regarding thermal energy storage: The different thermal energy storage techniques along with traditional materials that store thermal energy are canvassed in detail. A concise discussion regarding current status, leading groups, journals and the countries related to advance energy storage materials has also been discussed. But the poor performance of these materials is the main concern, which limits its applications. Thus, various advanced materials for efficient energy storage are proposed in literature, which are also discussed broadly for the first time in this book regarding comprehensive understanding for efficient thermal energy storage. The thermophysical properties of advanced materials and the role of these materials for thermal energy storage in different applications as buildings, solar energy, seawater desalination, cooling devices as well as photovoltaic thermal systems are also mentioned. The results suggest that advanced energy storage materials have a massive impact on heat transfer along energy storage capacity compared to conventional energy storage materials. Finally, emerging future research of advanced energy storage materials is also highlighted in this book, which will help in generating new insides for thermal energy storage development.

Keywords: Energy storage materials · PCM · Heat transfer · Thermophysical properties · Applications

Dhahran, Saudi Arabia — Hafiz Muhammad Ali
Taxila, Pakistan — Furqan Jamil
Sahiwal, Pakistan — Hamza Babar

Acknowledgements Hafiz Muhammad Ali is thankful for the support of King Fahd University of Petroleum and Minerals, Dhahran, Saudi Arabia.

Contents

Nomenclature

Abbreviations

a	Length of nanoparticle (m)
b	Diameter of nanoparticle (m^2)
C	Carbon atom
d	Mean diameter (m)
d_c	Diameters of core material
d_p	Diameters of micro/nano capsule
h	Convective heat transfer (W/m^2K)
h	Phase change enthalpy (KJ/kg)
k	Thermal conductivity (W/mK)
k_c	Thermal conductivity of core material
k_p	Thermal conductivity of micro/nano capsule
k_s	Thermal conductivity of shell material
m	Mass (kg)
T	Temperature (K)
V	Volume (m^3)
BN	Boron nitride
CO_2	Carbon dioxide
CNFs	Carbon nanofibers
CSP	Concentrated solar power
EC	Electrical conductivity
ENEA	Italian National Agency for New Technologies (Energy and Sustainable Economic Development)
G-NF	Graphene nanofibers
GA	Gum Arabic
GNP	Graphene NPs
GO	Graphene oxide
GJ	Gega Joule
H	Hydrogen
HT	Heat transfer

$H_{M,Micro\ PCM}$	Heat of fusion of microencapsulated PCM
$H_{M,PCM}$	Heat of fusion of PCM only
MWCNTs	Multi-wall carbon nanotube
MIT	Massachusetts Institute of Technology
MJ	Mega Joule
NPs	Nanoparticles
NFs	Nanofibers
Ni	Nickel
PA	Palmitic acid
PCM	Phase Change Material
PEG	Polyethylene glycol
PEG/Ti3C2Tx	Polyethylene glycol/MXene
PW	Paraffin Wax
PV/T	Photovoltaic/thermal
Q	Heat stored in material (KJ)
R	Encapsulation ratio
SEM	Scanning Electron Microscope
SWCNTs	Single-wall carbon nanotube
TC	Thermal conductivity
TES	Thermal energy storage
TEM	Transmission Electron Microscopy
UV-Vis-NIR	Ultraviolet-Visible-Near infrared region
W	Watt

Symbols

C_p	Specific heat (KJ/kgK)
ρ	Density ($\frac{\text{kg}}{\text{m}^3}$)

Chapter 1
Introduction

1.1 Summary

The chapter contains the background of thermal energy storage, which is one of the major factors for minimizing dependency on fossil fuels. If thermal energy is not stored, it will simply waste into the environment. The energy storage materials are employed as thermal energy storage system that has optimum thermal properties. The research in the last few decades exhibited that these materials have better advantages due to superior properties, which are discussed in this book.

1.2 Overview

The utilization of energy is escalating now-a-days with the enhancement in industrialization, better living standard and population. The energy consumption may be managed by fossil fuels and renewable energy. It is very clear that CO_2 is a major contributor towards climate change of the world and the big share of that CO_2 is evolved by fossil fuels burnt by power generation and transport sector. These two mentioned sectors release around 90% of CO_2 into the atmosphere [1].

The progressive development of CO_2 in atmosphere becomes the cause of temperature rise in the world and it melts polar ice caps. It turns causes the depletion of ozone layer, which also raises world temperature. So, it will not commercially favorable to use fossils fuel as energy resources in the future. In Europe, North America and Middle East; researchers started to think about renewable energy technology to avoid and control the problem of global warming caused by CO_2 emissions [2].

Nowadays, people want that they live in an ideal environment like's they want to live in cold climate in summers and hot climatic conditions in winters in their residential as well as working areas by using these so many invented technologies.

H. M. Ali et al., *Thermal Energy Storage*,
https://doi.org/10.1007/978-981-16-1131-5_1

This is also a factor that is increasing world demand. As world energy demand is increased hugely so dependency on conventional or non-renewable energy resources is also increased and potential of these resources decreasing rapidly.

The whole world is shifted to renewable resources of energy for their energy production. Solar energy is the most suitable and reasonable choice out of all renewable energy resources. According to IEA, 30% of electricity needs will be supplied through renewable energy by 2035. All new power generation systems will comprise around 60% clean sustainable energy technology by 2025 [3]. Nowadays, world is concentrating on renewable energy resources due to rapidly consumption of fossil fuels resources. There are many renewable energy resources accessible in world, among these resources; solar energy is the most eminent and clean energy resources, but these resources are still way to go.

In 2018, the world population increased by 1.13% yearly, which is a countable change, so the need for energy consumption by human being is also increasing [4]. Besides these, people of the world are entered such an age where everything became machinery [5]. With technology development tasks or works which once was done by human being are replaced by machine and equipment due to such increment of population, and introduction of so many machineries in our domestic industrial or any sector of life world energy consumption is also increasing drastically.

Enormous studies have been done to fill the energy supply for the efficient system [6, 7]. The storage of energy can be done by means of electrical, mechanical as well as thermal. The thermal energy storage may be in the form of sensible, latent, thermochemical or a combination of these. The temperature changes in materials during charging and discharging result in the storage of sensible heat. The latent heat storage deals with phase change behavior of materials. The latent heat is a superior way for thermal energy storage and results illustrated that latent heat showed storage density 5–10 higher than sensible heat [1, 8, 9]. The various methods for energy storage are sensible, latent and chemical in thermal energy storage systems as shown in Fig. 1.1.

The latent heat storage materials are recognized as phase change materials (PCM). The energy storage materials as PCMs study were initially investigated by Telkes and Raymond in 1940 [12] after PCMs were used for thermal energy storage, which released sensible and latent heat during 1973–1974 energy crisis [13]. The resources of fossil fuels are limited currently and green-house emissions make key issues for efficient energy utilization. The storage materials for thermal energy storage (TES) give an optimum solution to improve the performance of domestic as well as industrial applications [9, 10, 14–16]. The key benefits of using PCM for storing thermal energy are:

- Relatively constant temperature during charging and discharging.
- Storage capacity of maximum thermal energy related to the sensible energy storage in water.
- The backup generation unit of burner cycle; therefore, reducing harmful gases [17].

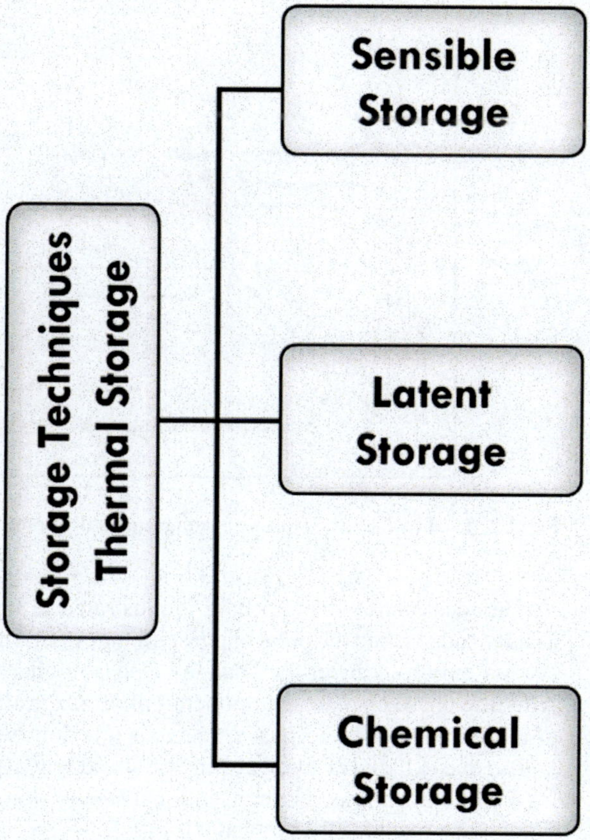

Fig. 1.1 Storage techniques and their classifications in thermal storage [10, 11]

The latent energy storage materials have a high value of heat of fusion to store or release large amount of heat. Different types of materials are under consideration containing organic compounds that included paraffin and fatty acids, inorganic materials as salts and their hydrates. These compounds have better heat of fusion and chemically stable [18] for thermal energy storage. The TES is broadly used in various applications and some of the benefits are [19]:

- The reliability of the system increased as the system works under more stable limits.
- The system generation capacity is enhanced, which is the most beneficial for demand sector.
- The cost for generation is optimized by thermal energy storage.

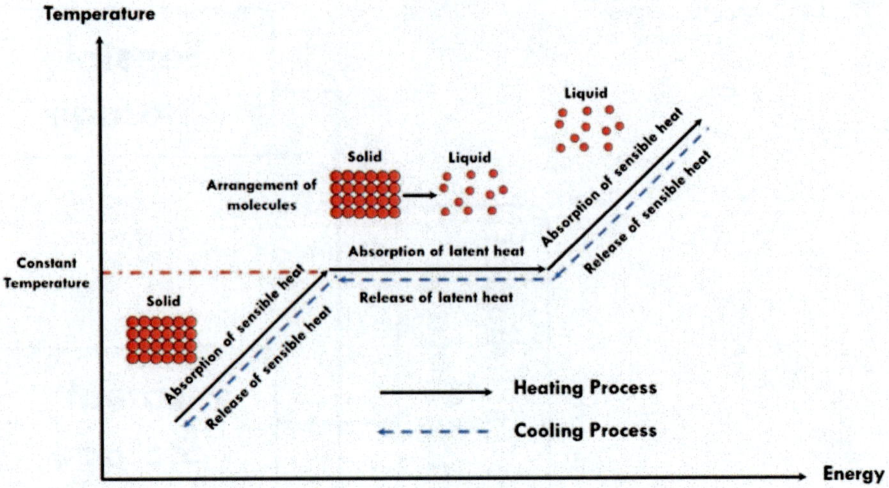

Fig. 1.2 Ideal case of PCM during heating and cooling process [20]

The ideal phase change curve is illustrated in Fig. 1.2 as temperature increases rapidly and reached the point of phase change of that material. At this, energy supplied is used only to convert the change of phase rather than increment of temperature. After the phase change, the temperature of material rises gradually. But the process of HT for conversion of phase change is too slow, which limits its usages.

But due to high HT rates, requirement and low TC of material make it a challenge for many applications. The conventional energy storage materials have a low value of TC and less rate of HT. The available energy storage materials do not meet the current requirements of the applications. The application of phase change material (PCMs) includes the techniques of storing thermal energy. Unfortunately, it is necessary to determine frequent impacts at the research and development phase before the important practical exercise of this technology. Some of the issues are as follows:

- Phase separation leads to inhomogeneous material distribution; hence the modification in the heat storage efficiency occurs; all properties vary in the space, and the storage procedure would be affected.
- Subcooling leads to change in phase at a low temperature than the probable one, i.e. the melting point; that means the production of energy will start at low temperature.
- Low thermal conductivity can cause problems as more time will require to attain desired temperature level in the whole material at the same heat flux, the temperature in the whole material will not be constant, and it will significantly vary from the heat source away.

The addition of nanoparticles (NPs) enhanced the TC of energy storage materials. The different types of nanostructured and nanofiber-based NPs integrated with conventional materials help in enhancing remarkable performance, which results in better HT. The carbon-based nanomaterials that include carbon nanofibers [21, 22] and carbon nanotubes [23, 24] are used for increment in performance of energy storage materials. The value of TC enhanced by about 40% by carbon nanotubes [24], graphite, graphene sheets increased TC by 140% [25] and carbon fibers up to 240% [26].

Various studies also have been done by different authors to enhance the properties of energy storage materials by micro-encapsulated or nano-encapsulated in which particles are encapsulated or using the combination of multiple energy storage materials called cascade PCM [27–32], which resulted in the increment of heat storage capacity. The encapsulation of PCM is useful to prevent the leakage in PCM during latent thermal energy storage that resulted in high HT and reduce reactivity with environment [33]. It is the process of coating individual particles or droplets with continuous film to produce capsule called microcapsule [34]. Encapsulated PCMs are made of two parts, namely, PCM as core material and polymer or inorganic shell material.

The molten salt-based nanofluids also enhanced the HT resulted in better performance [35–37]. These molten salts showed promising features for thermal energy storage as specific heat capacity and latent heat. Molten salts are considered as best storage media compared to conventional storage materials at high temperature up to 600 °C [38]. This case is also similar for cascade-type energy storage materials.

The energy storage materials have been extensively used in a wide range of applications as solar heating systems, building air conditioning, electronic cooling, waste heat recovery, textiles and a few more. The massive domain of energy storage materials is mostly used in solar applications. Sharma et al. [39] categorized PCMs as low and high temperature in solar energy applications. The low temperature included water, air, dryer heating and high temperature lead to solar cookers.

1.3 Work Conducted Around the World

The advanced energy storage materials have superior characteristics due to the most emerging heat transfer fluids and the most fascinating ones for future. This research interest of investigators around the world with the help of some stats of the last decade (2008–2018) is given below. According to statistics of "Web of Science", the 5,037 search results are found when search with the keyword "advance energy storage materials". The most energetic research group was "Chen Z", "Wang Y" and "Fang GY", which compiles the studied-on heat transfer behavior on energy storage materials.

Figures 1.3, 1.4, 1.5 and 1.6 illustrated the top 10 studies done during the last decade (2008–2018). The countries list from "Web of Science" indicated that China is the leading country where this study is more dominating. It is cleared that the number of publications is high in this country. Figures 1.3, 1.4, 1.5 and 1.6 show the

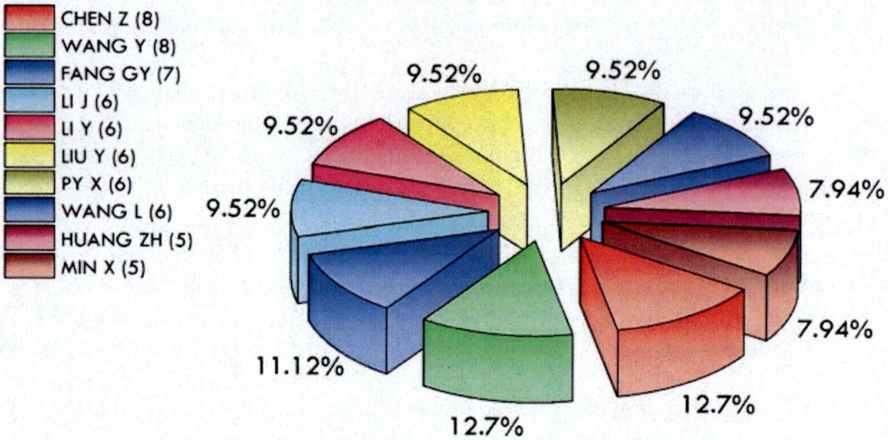

Fig. 1.3 Web of Science Stats of the top 10 authors published the work of the entire domain (searched with the keyword "energy storage materials" and refined with a word of "advanced")

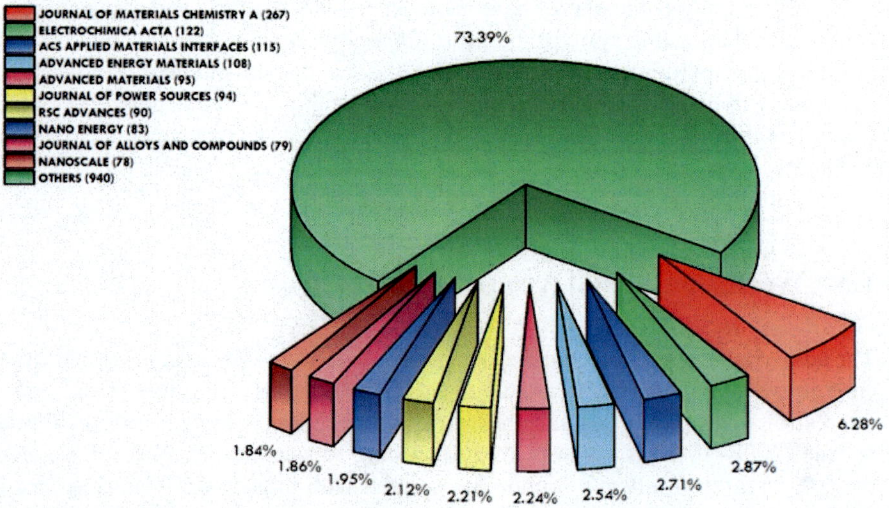

Fig. 1.4 Web of Science Stats of the top 10 journals published the work of the entire domain (searched with the keyword "energy storage materials" and refined with a word of "advanced")

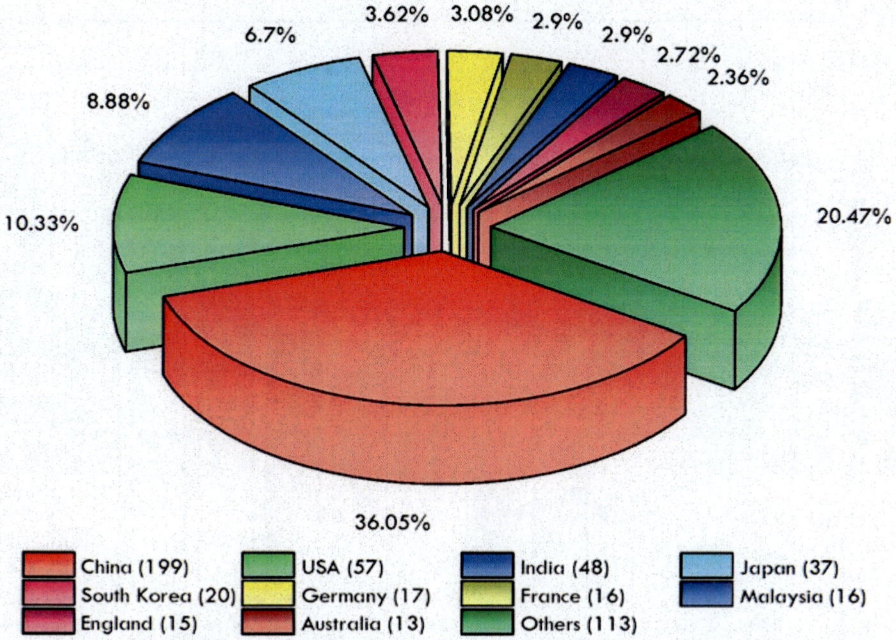

Fig. 1.5 Web of Science Stats of the top 10 countries published the work of the entire domain (searched with the keyword "energy storage materials" and refined with a word of "advanced")

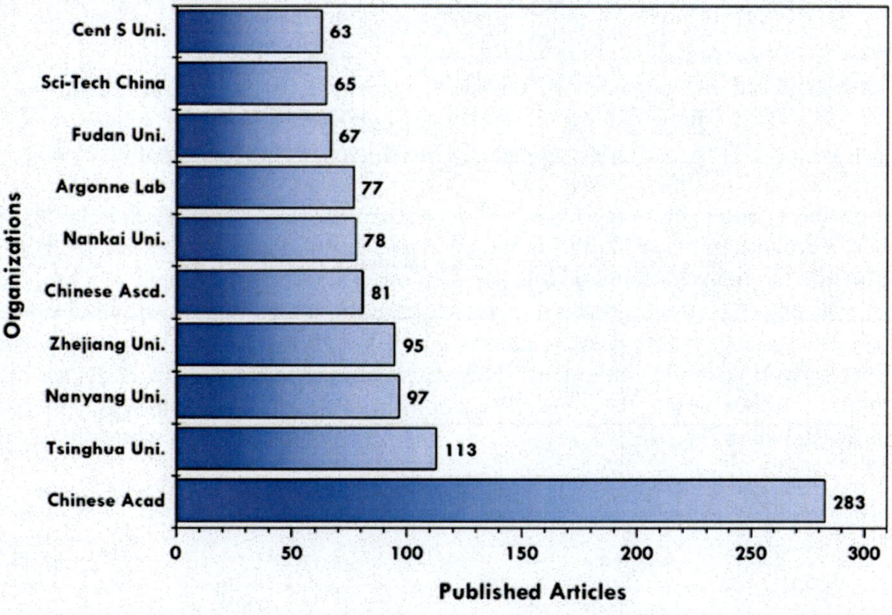

Fig. 1.6 Web of Science Stats of the top 10 organizations published the work of the entire domain (searched with the keyword "energy storage materials" and refined with a word of "advanced")

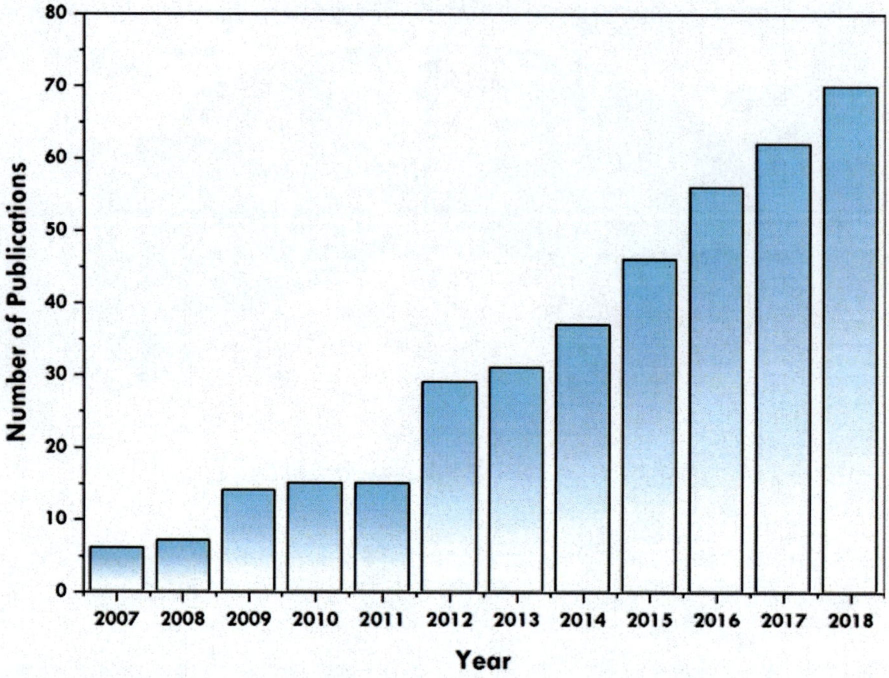

Fig. 1.7 Web of Science stats regarding a number of publications during (2008–2018)

graphical representation of the statistics of leading investigators, countries, journals and organizations that published the work on advanced energy storage materials.

Till to date, numerous experimental and numerical studies have been done to improve the HT of energy storage materials but definite conclusion is still not defined yet. Figure 1.7 illustrates the number of publications on advance energy storage materials during the year 2008–2018 as data taken from "Web of Science". It is cleared that number of publications increasing continuously, which indicated that the world is more focusing on this domain.

The objective of this book is to evaluate heat storage techniques, various advanced materials that can store energy better compared to already available energy storage materials. In addition, the major applications of energy storage materials are also under consideration in this book. The summary of the discussed study in this book is illustrated in Fig. 1.8.

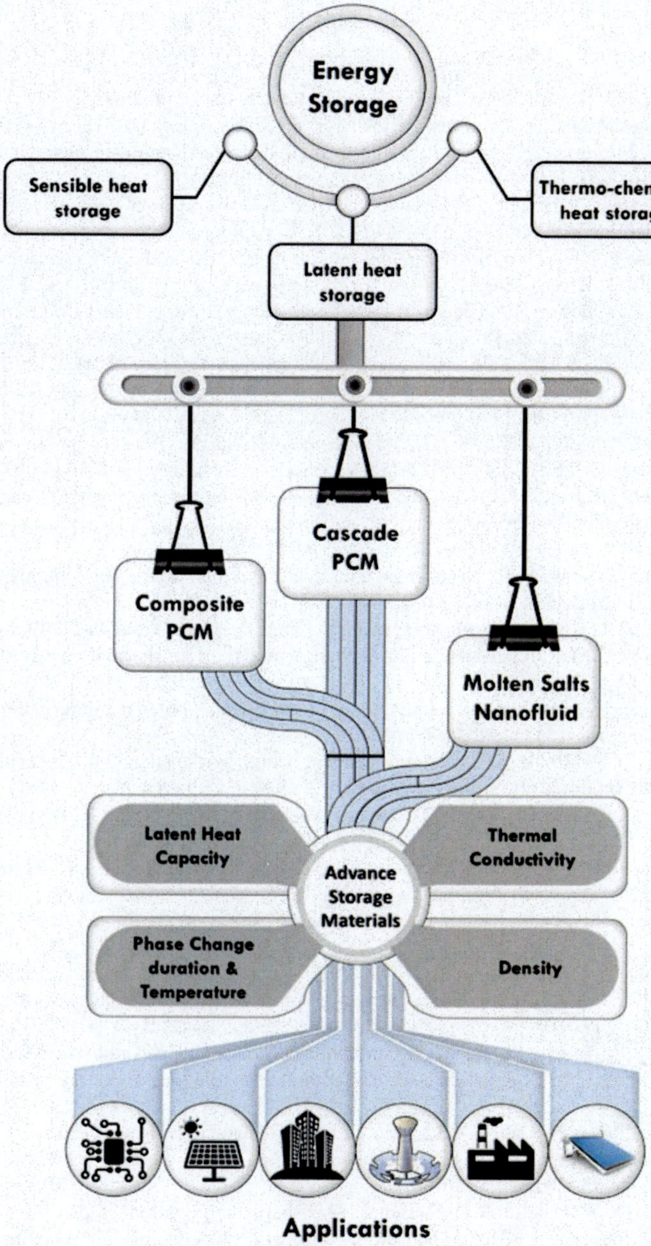

Fig. 1.8 Summarization of study discussed in the book

References

1. A. Sharma, V.V. Tyagi, C.R. Chen, D. Buddhi, Review on thermal energy storage with phase change materials and applications. Renew. Sustain. Energy Rev. **13**, 318–345 (2009)
2. S. Chu, A. Majumdar, Opportunities and challenges for a sustainable energy future. Nature **488**, 294–303 (2012). https://doi.org/10.1038/nature11475
3. A. Verbruggen, V. Lauber, Assessing the performance of renewable electricity support instruments. Energy Policy **45**, 635–644 (2012). https://doi.org/10.1016/j.enpol.2012.03.014
4. V. Lizunkov, Population of the world and regions as the principal energy consumer (2018)
5. A.D. Atrens, A. Atrens, Engineering in the modern world. Hum. Forces Eng. **1** (2018)
6. A. Abhat, Low temperature latent heat thermal energy storage: heat storage materials. Sol. Energy. **30**, 313–332 (1983)
7. A.F. Regin, S.C. Solanki, J.S. Saini, Heat transfer characteristics of thermal energy storage system using PCM capsules: a review. Renew. Sustain. Energy Rev. **12**, 2438–2458 (2008)
8. K. Kaygusuz, The viability of thermal energy storage. Energy Sources **21**, 745–755 (1999)
9. M.M. Farid, A.M. Khudhair, S.A.K. Razack, S. Al-Hallaj, A review on phase change energy storage: materials and applications. Energy Convers. Manag. **45**, 1597–1615 (2004)
10. B. Zalba, J.M. Marın, L.F. Cabeza, H. Mehling, Review on thermal energy storage with phase change: materials, heat transfer analysis and applications. Appl. Therm. Eng. **23**, 251–283 (2003)
11. N. Sarier, E. Onder, Organic phase change materials and their textile applications: an overview. Thermochim. Acta. **540**, 7–60 (2012)
12. M. Telkes, E. Raymond, Storing solar heat in chemicals. Heat. Vent. **46** (1949).
13. K. Pielichowska, K. Pielichowski, Phase change materials for thermal energy storage. Prog. Mater. Sci. **65**, 67–123 (2014)
14. M. Kenisarin, K. Mahkamov, Solar energy storage using phase change materials. Renew. Sustain. Energy Rev. **11**, 1913–1965 (2007)
15. H.P. Garg, S.C. Mullick, A.K. Bhargava, Latent heat or phase change thermal energy storage, in: Solar Thermal Energy Storage. (Springer, 1985), pp. 154–291
16. D. Fernandes, F. Pitié, G. Cáceres, J. Baeyens, Thermal energy storage: "How previous findings determine current research priorities." Energy **39**, 246–257 (2012)
17. E. Milisic, Modelling of energy storage using phase-change materials (PCM materials), 91 (2013)
18. S.M. Hasnain, Review on sustainable thermal energy storage technologies, part II: cool thermal storage. Energy Convers. Manag. **39**, 1139–1153 (1998)
19. M. Medrano, A. Gil, I. Martorell, X. Potau, L.F. Cabeza, State of the art on high-temperature thermal energy storage for power generation. Part 2—Case studies. Renew. Sustain. Energy Rev. **14**, 56–72 (2010).
20. K.Y. Leong, M.R.A. Rahman, B.A. Gurunathan, Nano-enhanced phase change materials: a review of thermo-physical properties, applications and challenges. J. Energy Storage. **21**, 18–31 (2019)
21. J. Li, L. He, T. Liu, X. Cao, H. Zhu, Preparation and characterization of PEG/SiO2 composites as shape-stabilized phase change materials for thermal energy storage. Sol. Energy Mater. Sol. Cells. **118**, 48–53 (2013)
22. R. Ehid, R.D. Weinstein, A.S. Fleischer, The shape stabilization of paraffin phase change material to reduce graphite nanofiber settling during the phase change process. Energy Convers. Manag. **57**, 60–67 (2012)
23. Y. Wang, B. Tang, S. Zhang, Single-walled carbon nanotube/phase change material composites: sunlight-driven, reversible, form-stable phase transitions for solar thermal energy storage. Adv. Funct. Mater. **23**, 4354–4360 (2013)
24. B. Li, S. Nie, Y. Hao, T. Liu, J. Zhu, S. Yan, Stearic-acid/carbon-nanotube composites with tailored shape-stabilized phase transitions and light–heat conversion for thermal energy storage. Energy Convers. Manag. **98**, 314–321 (2015)

25. F. Yavari, H.R. Fard, K. Pashayi, M.A. Rafiee, A. Zamiri, Z. Yu, R. Ozisik, T. Borca-Tasciuc, N. Koratkar, Enhanced thermal conductivity in a nanostructured phase change composite due to low concentration graphene additives. J. Phys. Chem. C. **115**, 8753–8758 (2011)
26. J. Wang, H. Xie, Z. Xin, Y. Li, C. Yin, Investigation on thermal properties of heat storage composites containing carbon fibers. J. Appl. Phys. **110**, 94302 (2011)
27. J. Wang, Y. Ouyang, G. Chen, Experimental study on charging processes of a cylindrical heat storage capsule employing multiple-phase-change materials. Int. J. Energy Res. **25**, 439–447 (2001)
28. J. Wang, G. Chen, F. Zheng, Study on phase change temperature distributions of composite PCMs in thermal energy storage systems. Int. J. Energy Res. **23**, 277–285 (1999)
29. H. Cui, X. Yuan, X. Hou, Thermal performance analysis for a heat receiver using multiple phase change materials. Appl. Therm. Eng. **23**, 2353–2361 (2003)
30. P. Charunyakorn, S. Sengupta, S.K. Roy, Forced convection heat transfer in microencapsulated phase change material slurries: flow in circular ducts. Int. J. Heat Mass Transf. **34**, 819–833 (1991)
31. S. Kuravi, K.M. Kota, J. Du, L.C. Chow, Numerical investigation of flow and heat transfer performance of nano-encapsulated phase change material slurry in microchannels. J. Heat Transfer. **131**, 62901 (2009)
32. R. Sabbah, M.M. Farid, S. Al-Hallaj, Micro-channel heat sink with slurry of water with micro-encapsulated phase change material: 3D-numerical study. Appl. Therm. Eng. **29**, 445–454 (2009)
33. G. Fang, Z. Chen, H. Li, Synthesis and properties of microencapsulated paraffin composites with SiO2 shell as thermal energy storage materials. Chem. Eng. J. **163**, 154–159 (2010)
34. V.V. Tyagi, S.C. Kaushik, S.K. Tyagi, T. Akiyama, Development of phase change materials based microencapsulated technology for buildings: a review. Renew. Sustain. Energy Rev. **15**, 1373–1391 (2011)
35. Z. Yang, S.V. Garimella, Thermal analysis of solar thermal energy storage in a molten-salt thermocline. Sol. Energy. **84**, 974–985 (2010)
36. Z. Yang, S.V. Garimella, Molten-salt thermal energy storage in thermoclines under different environmental boundary conditions. Appl. Energy. **87**, 3322–3329 (2010)
37. R.W. Bradshaw, N.P. Siegel, Molten nitrate salt development for thermal energy storage in parabolic trough solar power systems, in: ASME 2008 2nd International Conference on Energy Sustainable Collocated with Heat Transfer Fluids Engineering, 3rd Energy Nanotechnology Conference. American Society of Mechanical Engineers (2008), pp. 631–637
38. T.R. Shah, H.M. Ali, Applications of hybrid nanofluids in solar energy, practical limitations and challenges: A critical review. Sol. Energy. **183**, 173–203 (2019)
39. R.K. Sharma, P. Ganesan, V.V. Tyagi, H.S.C. Metselaar, S.C. Sandaran, Developments in organic solid–liquid phase change materials and their applications in thermal energy storage. Energy Convers. Manag. **95**, 193–228 (2015)

Chapter 2
Thermal Energy Storage System

2.1 Summary

This chapter includes various thermal energy storage systems and different conventional energy storage materials used in that system. The conventional energy storage materials are for low and medium temperature range applications and are tested under these conditions.

2.2 Classification of the Thermal Energy Storage System

The thermal energy storage system can be classified based on various categories. Based on temperature range, it can be divided as low-temperature thermal energy storage (LTTES) system and high-temperature thermal energy storage (HTTES) system [1, 2]. For LTTES, the temperature is below 200 °C while for HTTES, temperature feasibly is above 200 °C, which mostly use in waste heat recovery systems. If according to time length, it can be categorized as "short-term" and "long-term" system. If based on energy storage media, then it may be categorized by sensible as well as latent heat storage system. Various classifications of thermal energy storage system are illustrated in Fig. 2.1.

2.3 Thermal Energy Storage Based on Storage Media

2.3.1 Sensible Heat Storage

In this mode, the heat is stored by temperature change of material only. The basic principle is that heat is absorbed by storage materials using conventional heat transfer

H. M. Ali et al., *Thermal Energy Storage*,
https://doi.org/10.1007/978-981-16-1131-5_2

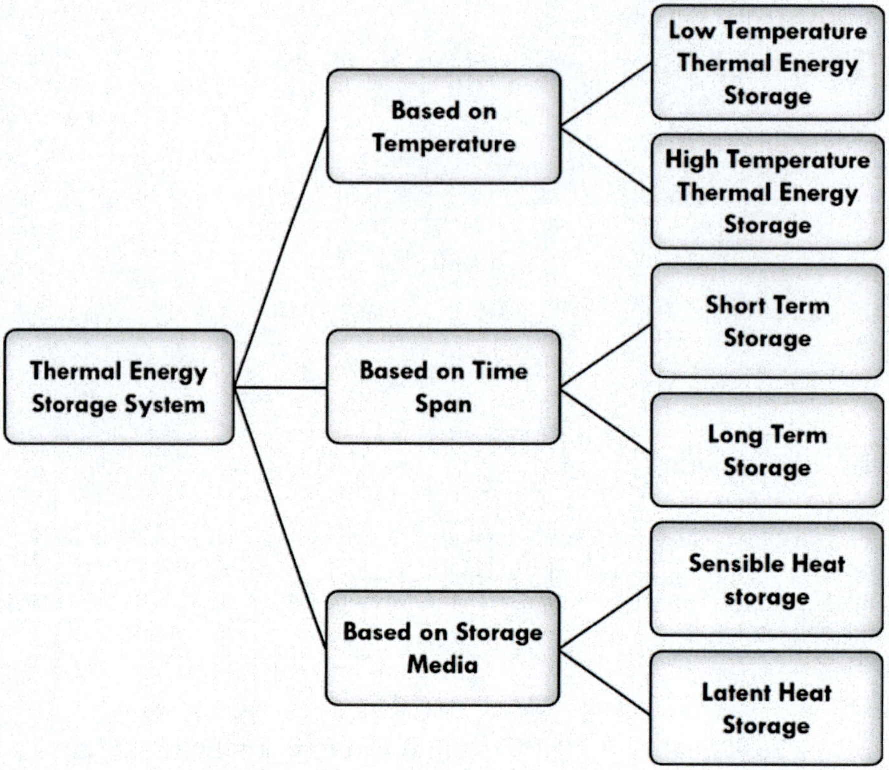

Fig. 2.1 Classification of thermal energy storage system

phenomena either conduction, convection or radiations, which also releases heat during night with same heat transfer phenomena.

The sensible heat storage system includes hot as well as cold-water tanks and underground storage systems. The underground thermal storage refers to aquifer or borehole thermal energy storage, which exchanges heat by using a heat exchanger [3]. The sensible heat storage heat (Q) may be written as [4]:

$$Q = \rho \times V \times C_p \times \Delta T \tag{2.1}$$

where $\rho \times V$ is the density and volume of heat storage medium, C_p is the specific heat and ΔT is the temperature difference between initial and final temperature The sensible heat storage system consists of storage medium, container and inlet/outlet devices. The storage medium must be such material that it should continue to hold storage material, retrain thermal energy losses and maintain temperature difference across the storage medium [5].

The various sensible energy storage materials are liquid or solid media. The liquid media includes water, thermal oils, molten salts and solid media appends metals, earth

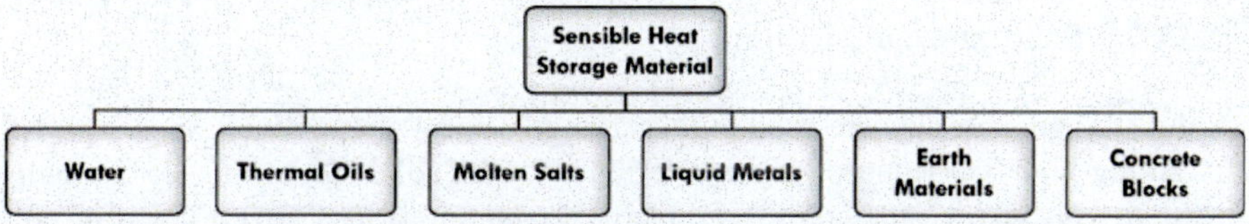

Fig. 2.2 Various sensible energy-storing materials

materials, concrete blocks and cast ceramic as depicted in Fig. 2.2. The solid media are placed in packed bed for good heat exchange [6]. For good heating purposes, rocks bin or water tanks are extensively used [7]. The thermal oils and molten salts are extensively used in solar towers and parabolic trough collectors applications [8].

2.3.1.1 Liquid Storage Medium

The liquid storage materials are used where heat transportation can be required, or it may be circulated through system, which is called active system. The heating of fluid causes density difference results in creating the bouncing effect, which subsequently producing temperature gradient [9].

In liquid media, water is the most available storage fluids for low-temperature applications within the range 25–90 °C, since it can be circulated so it may be used as active system in heat transfer or thermal energy storage material. The water is non-toxic, cheap, abundant and high heat storage capacity of water is 250 MJ/m^3 at 60 °C [10]. The water sensible heat can be stored in tanks, aquifer or solar ponds [11]. But the main disadvantages of using water are [12]:

- System leakage and corrosion
- Due to high vapor pressure, high cost insulation is required
- For addition or rejection of heat large swing is required.

Thermal oils have better heat transfer quality, which are organic fluids having colorless clear liquids. The vapor pressure of thermal oil is lower than water, which reduces the overall cost of system. The corrosion occurred in pipes when oil has high temperature, which resulted in oxidation of air [13].

The metals that have low melting point but high boiling point are best suited for sensible heat storage. The thermal efficiency is high up to 50 % if working condition is on high operating temperature. The sodium has thermal conductivity 64.9 W/mK and heat transfer coefficient 18,000–28,500 W/m^{2}K [14]. But liquid metals have some disadvantages:

- Liquid metals are expensive
- Corrosion may be occurred
- Some metals are toxic and flammable.

2.3.1.2 Solid Storage Medium

Solid storage materials are used in that storage having temperature range 200–1200 °C along high TC value of 1–40 W/mK [15]. The solid storage materials include naturally occurring metals, rocks, concrete, sand and bricks [16]. They are specifically useful when they are used in packed bed structure [17]. In this method, the fluid is in direct contact between the packed bed and exchange media.

The rocks are extensively used due to the low cost with size range from 1 to 5 cm [2]. The heat transfer area enhanced by using rock resulted in less heat loss. The low cost, locally available, heat capacity, density and thermal conductivity are critical parameters while selecting a rock. By using the rock, there are many advantages as.

- Rocks are non-toxic and non-flammable
- It behaves as heat storage medium and heat transfer medium
- Rocks are inexpensive.

Concrete blocks are also used for solid heat storage medium. The mechanical properties of concrete show good results, which help in reducing cost and also the concrete is less expensive, non-flammable and non-toxic. The sand, gravel and cement must be combined to form concrete. The sand is mainly silica, gravels are type of rock and cement acts as binder. Concrete was under consideration, which was based on calcium alumina cement within temperature range 290–550 °C and suitable up to 550 °C for thermal energy storage [18]. The summary of various sensible heat storage materials used by different researchers is given in Table 2.1.

2.3.2 *Latent Heat Storage*

The energy storage capacity for sensible heat storage materials is low, which results in latent heat storage materials. Thus, thermal energy is stored and released by altering the phase without altering the temperature in this mode. The thermal latent heat storage (Q) may be written as [28]:

$$Q = m \times \Delta h \tag{2.2}$$

where m is the mass of heat storage material and Δh is the phase change enthalpy. The latent heat storage method may be from solid–solid, solid–liquid, solid–gas, liquid–gas. In solid–solid phase, heat is stored in material and there is little heat addition and small volume change will be occurred. The solid–liquid can store large heat over a small temperature range. The solid–gas and liquid–gas have high latent heat values but high volume changes are not suitable for many applications [29]. The latent heat storage materials can be considered phase change materials, which may be divided into organic and inorganic, eutectic. The organic materials contain paraffin and non-paraffin materials. The non-paraffins include various materials such

Table 2.1 Sensible heat storage materials studied by various authors

Material	Type	Thermal conductivity (W/mK)	Density (kg/m^3)	Specific heat capacity (KJ/kgK)	Temperature Range °C	Cost	References
Ceramic brick	Solid	0.73	1800	0.92	–	36–64 €/m^3	[10]
Therminol VP-1	Liquid	0.11	1068 @ 40 °C	2.075	–	25,000 €/ton	[13]
Vegetable oil	Liquid	0.11	926 @ 40 °C	2.509	–	835 €/ton	[13]
Stone, granite	Solid	1.3–3.8	2640	0.82	≤ 160	–	[19]
Wood	Solid	0.12	450	1.6	–	720 €/m^3	[20]
Octane	Liquid	0.134	704	2.4	≤ 126	–	[21]
Asphalt sheet	Solid	1.2	2300	1.7	–	56 €/m^3	[22]
Water	Liquid	0.63	990 @ 40 °C	4.19	0–100	1.6 €/m^3	[22]
Steel Slab	Solid	50	7800	0.502	20–70	–	[22]
Hitec XL, solar salt KNO_3-$Ca(NO_3)_2$ (7–45-48)	Liquid	0.519	1992	1.44	–	–	[23, 24]
Ethanol	Liquid	0.171	790	2.4	≤ 78	–	[25]
Quartz	Solid	2.0	2210–2770	0.8–0.9	–	–	[26]
Granite	Solid	2.8	2530–2620	0.6–1.2	–	–	[26]
Sodium (Na)	Liquid	64.9	1042	1.3	–	–	[27]

Table 2.2 Various latent heat storage materials [44]

Materials	Heat of fusion (KJ/kg)	Melting Temperature (°C)	Freezing point range (°C)	Group[a]
6106[b]	189	–	42–44	I
6403[b]	189	–	62–64	I
6499[b]	189	–	66–68	I
Formic acid	247	7.8	–	III
Phenol	120	120	–	III
Cyanamide	209	44	–	II
Acetic acid	184	16.7	–	I
Palmatic acid	163	55	–	I
Methyl fumarate	242	102	–	I
$MgSO_4.7H_2O$	202	48.5	–	II
$Ba(OH)_2.8H_2O$	265	78	–	II
Lead	22.4	327	–	–
Gold	67	1063	–	–
Heptane	140	−90.5	–	–
Coconut oil	103	24	–	–
Paraffin wax	147	46	–	–
Hexane	152	−95	–	–
Ethylene–glycol	181	−12.8	–	–
Dodecanese	216	−25.8	–	–
Aluminum	321	658	–	–
Water	334	0	–	–
Ammonia	339	−78	–	–

[a] Group I, most promising group; Group II, promising group; Group III, less promising group,
[b] Manufacturer of technical Grade Paraffin's 6106,6403,6499 Ter Hell Paraffin Hamburg, FRG

as ester, fatty acids, alcohol and glycol. The inorganic materials include salts, salt hydrates, salt eutectic, metal and alloys. The solid–solid media included polyurethane and polybutadiene materials. The various solid–liquid and solid–solid phase change materials are summarized in Fig. 2.3. The summery of various latent heat storage materials is given in Table 2.2.

Few of the characteristics of the phase change materials are:

- Non-flammable
- Non-toxicity
- Relatively low cost
- High availability
- High density
- Chemical stability
- High specific heat capacity

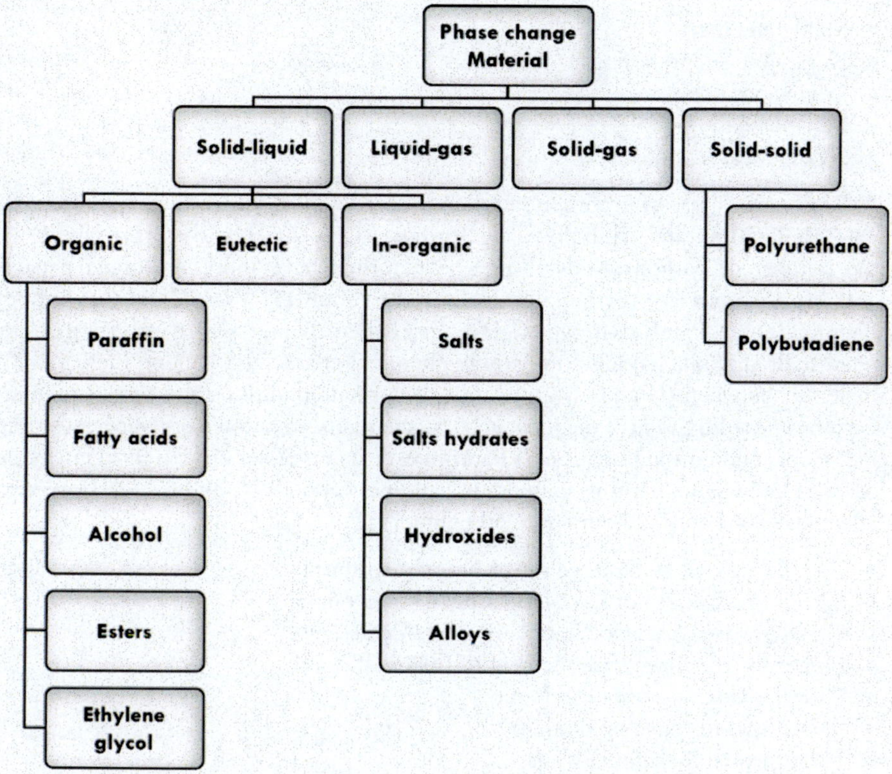

Fig. 2.3 Types of latent heat storage materials

- Heat Latent Capacitance (Storage Capacity)
- Non-corrosive if used in some structural materials
- Minimum volume changes during the phase change
- No degradation for repeated multiple numbers of cycles
- Favorable thermal conductivity assisting in heat absorption as well as release in thermal storage system
- As phase transition takes place within the desirable operating temperature range so that the bulk of input heat is stored.

2.3.2.1 Organic Materials

The organic materials lie between solid–liquid interface and organic materials can be categorized by paraffin organic materials as well as inorganic paraffin materials [6]. The paraffin materials can be considered as good phase change materials because [30–32]:

- High heat of fusion

- Stable and inert
- Negligible supercooling
- Low vapor pressure
- Self-nucleating
- Available at low cost.

The various paraffin compounds have been studied by different authors [30, 33–35] Paraffins are symbolized by formula ($CH_3-(CH_2)_{n-2}-CH_3$) where n is the number of carbon atoms [36]. The paraffins have thermal storage capacity 150 MJ/m^3 and low thermal conductivity of about 0.2 W/m°C but they are less flammable [37]. Paraffins of type C_nH_{2n+2} are called alkanes, which is a group of saturated hydrocarbons having comparable chemical characters. Among these paraffin materials between C_5 and C_{15} are fluids in normal conditions and rest of them are waxen-type solids. One of the most general and readily available organic heat storage PCMs for economic applications is Paraffin wax, it comprises on straight chain hydrocarbons and have a melting temperature run between 23 °C and 67 °C. Here some advantages of Paraffin are mentioned below [38]:

- Paraffin wax exists in large range of temperature.
- Paraffin wax does not have tendency to super cool.
- Paraffin waxes are also chemically remained stable.
- Paraffin waxes have catholic heat of fusion.
- Paraffin waxes do not detach out.
- Paraffin waxes have no damage.
- Paraffin wax does not corrode.

The non-paraffin materials include easter, fatty acids, alcohol, glycol. Easter may be represented by R-COO-R^1 where R and R^1 are alkyl groups and can be obtained by carboxylic acid and alcohol by a process called esterification [39].

A review on fatty acids was studied by Yuan et al. [40] for phase change materials. The important were lauric acid, myristic acid, palmitic acid and stearic acid but they have thermal conductivity from 0.15 to 0.17 W/mK. Fatty acids have general formula R-COOH where R is alkyl group and they may be saturated or unsaturated. Fatty acids [41, 42] have some properties as:

- High heat of fusion
- Reproducible melting and freezing behavior
- Freezing with minor supercooling.

Alcohol has high melting and latent heat and their phase change temperature is best for solar heaters and waste heat recovery for temperature range 90–250 °C [43]. Glycol has also been used for storage as in case of polyethylene glycol (PEG), which has general formula $H-(O-CH_2-CH_2)_n-H$. The temperature difference between PEG and other organic is about 30–40 °C [6].

2.3.2.2 Inorganic Materials

The inorganic materials can be classified into various categories such as salts, salt hydrates, salt eutectic and metal alloys. Salt hydrates consist of salt and water to form a crystalline when solidified. It may be utilized in form of separately or eutectic mixture [45]. The various types of salt hydrates were investigated by different authors [46–48]. Salt hydrates can be represented by general form as "anhydrous salt.nH_2O". Salt hydrates are considered having better heat storage capacity due to [49–51]:

- Easily available with low cost
- Melting point is high
- TC has large value.

Marks [52] studied the stability of Glauber's salt by using borax as nucleate agent and attapulgite clay as a thickener. For measuring thermal storage capacity, the calorimeter measurements were taken, and results showed that thermal capacity of pure salt decreased rapidly and Glauber's salt was not suitable for long duration of latent heat storage. Calcium chloride hexahydrate ($CaCl_2.6H_2O$) is a renowned salt hydrate and stability was considered by Kimura and Kai [53] with little increase in water concentration. The phase separation occurred as ($CaCl_2.4H_2O$) is formed, which tends to settle at bottom. By adding sodium chloride that has good nucleate ability resulted in no phase separation.

Metal and alloys diminish the problems that occurred in inorganic salts as low conductivity and high-volume change. The metals and alloys have the highest heat storage capacity, highest thermal conductivity and melting temperature above 300 °C [54]. But they have low value of heat storage per unit weight, which results in weight problem [55]. Birchenal and Telkes [56] were first analyzed the metals to store energy using heat of fusion of metals and afterward various authors [57–59] studied alloys for thermal energy storage. Birchenall and Riechman [59] investigated metal eutectic alloys to store thermal energy and thermal properties were determined. The experimental results indicated that the error was 4% for heat of fusion and 3 K for melting temperature and the eutectic was found to be best in terms of properties.

Li et al. [60] investigated aluminum–silicon alloy used as PCM. The authors concluded that aluminum silicon alloys are stable during heating and cooling cycles. Kotze et al. [61] studied eutectic alloy of aluminum and silicon as metallic PCM. The thermal performance of spherical capsules that include metal/metal alloys was studied by Akiyama et al. [62] for heat storage and release process. The results showed that these materials have constant temperature heat transfer fluid as they have high value of TC.

The aluminum silicon alloy was also studied by Wang et al. [63] using differential scanning calorimeter. It was selected for testing in electric heater further as high heat of fusion was high. The thermal stability of Al-Zn alloy was investigated by Sun et al. [64], which was used for latent heat storage. The results illustrated that temperature change was 3.06–5.3 °C and latent heat of fusion decreased by 10.98% after 1000 cycles, which showed positive reliability. Risueno et al. [65] developed light–based metal alloys Mg–Zn–Al for storage applications. The authors recommended that among various studied composition, the Mg–Zn–Al having composition of (70/24.9/5.1 wt%) was best for regarding properties. The various metal alloys under-considered by different authors are tabulated in Table 2.3. The metal and alloys are some disadvantages due to:

- Strain occurred due to thermal expansion coefficient
- Supercooling in pure metals
- They are complex in operation.

The properties of the discussed conventional energy storage materials are summarized in Fig. 2.4. The organic type of energy storage materials has higher phase change temperature, stable and recyclability. The inorganic PCMs have excellent

Table 2.3 Metal Alloys studied by various authors

Compound	Melting Temperature (°C)	Heat of fusion (KJ/kg)	References
Mg-Zn (48/52 wt %)	340	180	[57]
Al–Mg-Zn (59/33/6 wt %)	443	310	[57]
Mg-Cu–Zn (60/25/15 wt %)	452	254	[57]
Mg-Cu-Ca (52/25/23 wt %)	453	184	[57]
Zn-Al (96/4 wt %)	381	138	[62]
Mg–Al (34.65/65.35 wt %)	497	285	[62]
Al-Si-Cu-Mg (60.8/33.2 wt %)	506	365	[62]
Al-Si (12/86 wt %)	576	560	[64]
Al–Mg-Zn (60/34/6 wt %)	450.3	329.1	[66]

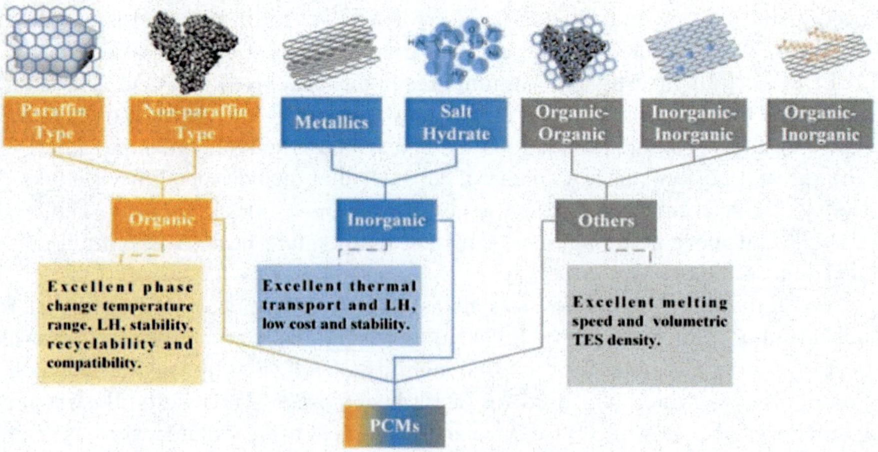

Fig. 2.4 Summarization of properties for conventional energy storage materials [30]

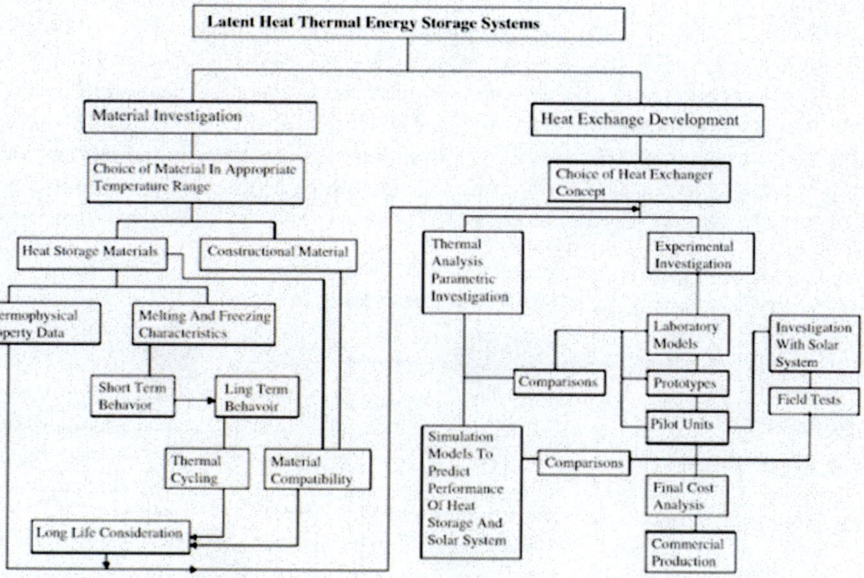

Fig. 2.5 Flow diagram for different stages in the development of thermal storage system [30]

thermal properties and latent heat and also have low cost and better stability, while the other types of PCMs contained excellent melting and thermal energy storage density.

The development of latent heat thermal energy storage based on three advanced knowledge of subjects includes phase change materials, material containers and heat exchanger. The technical options for storing low-temperature thermal energy are illustrated in Fig. 2.5.

2.3.3 *Thermochemical Heat Storage System*

The sorption system (adsorption and absorption) is chemical heat storage. In adsorption, the adsorptive gradually gathers on adsorption surface and forms a molecular layer whereas absorption occurs when substance is spread into liquid or solid substance and creates solution [67].

The chemical heat storage system stores and releases heat by the formation or breaking chemical bonds. The most advantage of using this method is negligible loss of energy during the reaction at ambient temperature. Thermochemical energy storage has higher storage density than other thermal energy storage methods. The basic principle for thermochemical reaction can be expressed by following reversible reaction [67]:

$$C + \text{heat} \rightleftarrows A + B \quad (2.3)$$

In this reaction, the thermochemical material C absorbed heat and converted in to two components A and B, which can be stored separately. The same reaction in reverse direction by releasing the heat and storage capacity is the amount of heat stored during material C formation. The main three processes are charging, storing and discharging as represented in Fig. 2.6.

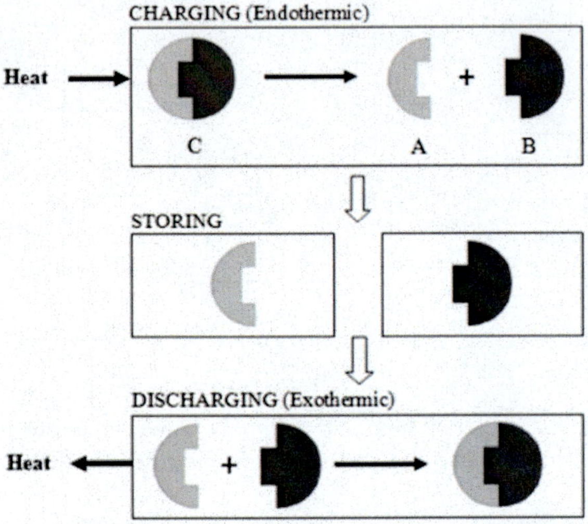

Fig. 2.6 Various processes in thermochemical energy storage cycle [67]

Fig. 2.7 Apparatus for testing of chemical energy storage [68]

The apparatus used by Yan and Zhao [68] for chemical energy storage is shown in Fig. 2.6. The main components of the apparatus were reactor and steam generator. The storage material was in reactor and as pressure decreased in reactor then temperature also decreased, and pressure was maintained by vacuum pump, which resulted in controlling the charging and discharging process. The chemical heat storage decomposed during charging and it can be stored for long duration, which made it feasible (Fig. 2.7).

The thermochemical storage system can be sorption system in which a small amount of energy having low temperature is produced. Also, lower activation energy is required to start reaction, which is used in low-temperature applications. The various sorption methods are liquid absorption, solid absorption, chemical reaction and by using composite materials.

The sorption system may be categorized by open and close system. The open system is open to environment and sorbate comes directly from air or humidifier. In closed cycle, the heat stored as sensibly and extracted [69]. The performance of thermochemical storage depends upon the following factors

- Type of thermochemical material
- Reactor design
- Process design
- Ambient conditions.

Zondag et al. [70] made protype for thermochemical heat storage system at Energy Research Center in Netherland. The system was packed bed with 17 dm^3 of magnesium chloride ($MgCl_2.2H_2O$) sorption material. The result indicated that 0.5 GJ/m^3 energy density obtained from such system and 150 W power produced by sorption material for 40 h with airflow of 410 L/min but only 50 W transferred due to heat loss (Fig. 2.8). Abedin and Rosen [71] studied open and close cycle with the aid of energy and exergy analysis for determining the performance and to investigate charging behavior. The energy and exergy efficiencies of closed system were 50% and 9% and for open system, it was 68% and 23% respectively. The energy density measurement is a useful technique to estimate the efficiency of storage materials. The energy density measurement showed that metal oxides, metal chlorides and metal hydrides are the best materials for thermochemical energy storage [72] and energy

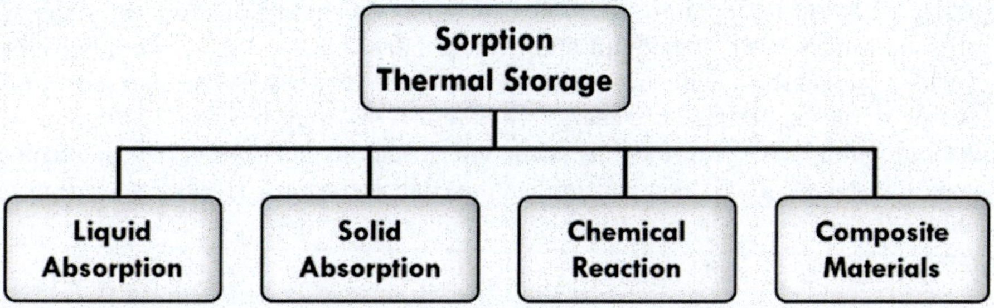

Fig. 2.8 Various sorption thermal methods

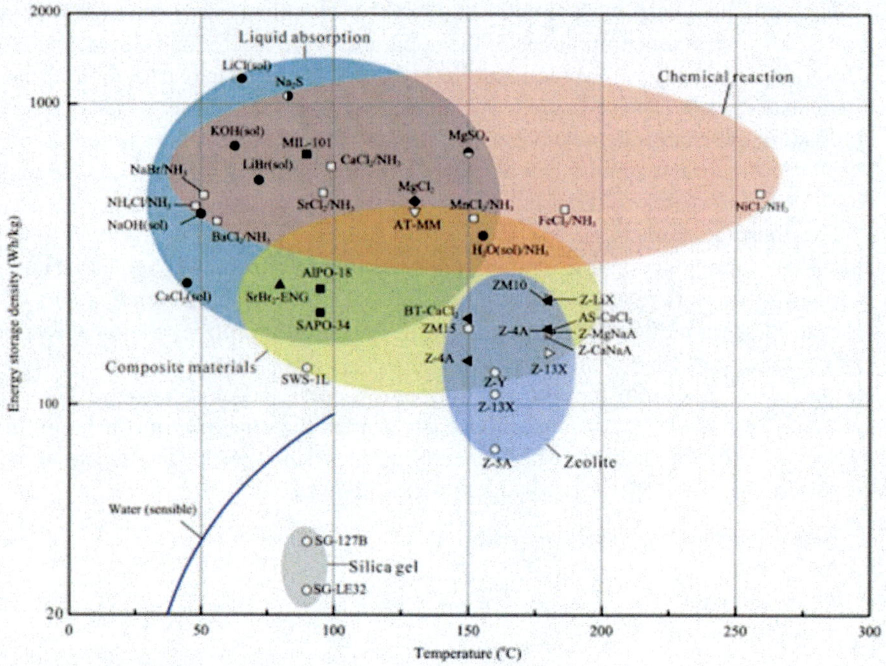

Fig. 2.9 Different materials with energy storage densities [72]

density of various materials is illustrated in Fig. 2.9 to make an estimation regarding efficiency.

The summary of various conventional energy storage materials is discussed in this chapter. This includes sensible, latent and thermochemical storage materials. The sensible heat storage materials have less capacity to store energy. The latent heat storage materials mainly discussed that include organic type of energy storage materials, inorganic PCMs in which some have better thermal properties while other types of PCMs have good thermal energy storage density. The corrosion of salts and salts hydrates for storage needs further investigation for the compatibility storage media as well as storage system that has a direct impact on stability and tenure for storage. For low temperature, sensible heat storage materials include water, thermal oils and sand while for medium temperature latent heat storage materials include organic, inorganic and eutectic materials but they are very costly. The thermochemical energy storage materials store energy by chemical reactions process but their storage is still low, and a lot of research is needed to enrich this field. Also, thermochemical heat storage has limited practical applications due to currently laboratory stage usage only.

References

1. S.M. Hasnain, Review on sustainable thermal energy storage technologies, Part II: cool thermal storage. Energy Convers. Manag. **39**, 1139–1153 (1998)
2. S.M. Hasnain, Review on sustainable thermal energy storage technologies, part I: heat storage materials and techniques. Energy Convers. Manag. **39**, 1127–1138 (1998)
3. J. Baker, New technology and possible advances in energy storage. Energy Policy. **36**, 4368–4373 (2008)
4. H.P. Garg, S.C. Mullick, A.K. Bhargava, Sensible heat storage, in *Solar Thermal Energy Storage* (Springer, 1985), pp. 82–153
5. T. Kousksou, P. Bruel, A. Jamil, T. El Rhafiki, Y. Zeraouli, Energy storage: applications and challenges. Sol. Energy Mater. Sol. Cells. **120**, 59–80 (2014)
6. G. Alva, Y. Lin, G. Fang, An overview of thermal energy storage systems. Energy **144**, 341–378 (2018)
7. K. Pielichowska, K. Pielichowski, Phase change materials for thermal energy storage. Prog. Mater. Sci. **65**, 67–123 (2014)
8. H. Zhang, J. Baeyens, G. Caceres, J. Degreve, Y. Lv, Thermal energy storage: recent developments and practical aspects. Prog. Energy Combust. Sci. **53**, 1–40 (2016)
9. A. Gil, M. Medrano, I. Martorell, A. Lázaro, P. Dolado, B. Zalba, L.F. Cabeza, State of the art on high temperature thermal energy storage for power generation. Part 1—concepts, materials and modellization. Renew. Sustain. Energy Rev. **14**, 31–55
10. P. Tatsidjodoung, N. Le Pierrès, L. Luo, A review of potential materials for thermal energy storage in building applications. Renew. Sustain. Energy Rev. **18**, 327–349 (2013)
11. Z.J. Zhai, M.L.L. Abarr, S.N.J. AL-Saadi, P. Yate, Energy storage for residential buildings: review and advances, in: *Proceedings on APEC Conference Low-Carbon Towns Physical Energy Storage* (2013)
12. C. Wyman, J. Castle, F. Kreith, A review of collector and energy storage technology for intermediate temperature applications. Sol. Energy. **24**, 517–540 (1980)
13. E.S. Kenda, K.E. N'Tsoukpoe, I.W.K. Ouédraogo, Y. Coulibaly, X. Py, F.M.A.W. Ouédraogo, Jatropha curcas crude oil as heat transfer fluid or thermal energy storage material for concentrating solar power plants. Energy Sustain. Dev. **40**, 59–67 (2017)
14. H. Benoit, L. Spreafico, D. Gauthier, G. Flamant, Review of heat transfer fluids in tube-receivers used in concentrating solar thermal systems: Properties and heat transfer coefficients. Renew. Sustain. Energy Rev. **55**, 298–315 (2016)
15. U. Pelay, L. Luo, Y. Fan, D. Stitou, M. Rood, Thermal energy storage systems for concentrated solar power plants. Renew. Sustain. Energy Rev. **79**, 82–100 (2017)
16. M.M. Khin, Encapsulation of phase change materials (PCMs) for heat storage (BE, Yangon Technological University) (2003)
17. H. Singh, R.P. Saini, J.S. Saini, A review on packed bed solar energy storage systems. Renew. Sustain. Energy Rev. **14**, 1059–1069 (2010)
18. M.C. Alonso, J. Vera-Agullo, L. Guerreiro, V. Flor-Laguna, M. Sanchez, M. Collares-Pereira, Calcium aluminate based cement for concrete to be used as thermal energy storage in solar thermal electricity plants. Cem. Concr. Res. **82**, 74–86 (2016)
19. O.E. Ataer, Storage of thermal energy and Eolss (2009)
20. A. de Gracia, L.F. Cabeza, Phase change materials and thermal energy storage for buildings. Energy Build. **103**, 414–419 (2015)
21. H. Watanabe, D.J. Seong, The thermal conductivity and thermal diffusivity of liquid n-alkanes: C n H 2n+ 2 (n= 5 to 10) and toluene. Int. J. Thermophys. **23**, 337–356 (2002)
22. H. Asan, Y.S. Sancaktar, Effects of wall's thermophysical properties on time lag and decrement factor. Energy Build. **28**, 159–166 (1998)
23. E. González-Roubaud, D. Pérez-Osorio, C. Prieto, Review of commercial thermal energy storage in concentrated solar power plants: steam vs. molten salts. Renew. Sustain. Energy Rev. **80**, 133–148 (2017)

24. G. Alva, L. Liu, X. Huang, G. Fang, Thermal energy storage materials and systems for solar energy applications. Renew. Sustain. Energy Rev. **68**, 693–706 (2017)
25. J. Wilson, Thermal diffusivity. Electron. Cool. **13**, 7 (2007)
26. H. Grirate, N. Zari, I. Elamrani, R. Couturier, A. Elmchaouri, S. Belcadi, P. Tochon, Characterization of several Moroccan rocks used as filler material for thermal energy storage in CSP power plants. Energy Procedia. **49**, 810–819 (2014)
27. J. Pacio, A. Fritsch, C. Singer, R. Uhlig, Liquid metals as efficient coolants for high-intensity point-focus receivers: implications to the design and performance of next-generation CSP systems. Energy Procedia. **49**, 647–655 (2014)
28. L.F. Cabeza, *Advances in Thermal Energy Storage Systems: Methods and Applications*. Elsevier (2014)
29. S.M. Hasnain, Energex'93, in: *Energy Conference* (1993)
30. A. Abhat, Low temperature latent heat thermal energy storage: heat storage materials. Sol. Energy. **30**, 313–332 (1983)
31. G.A. Lane, Solar heat storage: latent heat materials (1983)
32. B.M. Gibbs, S.M. Hasnain, *DSC Study of Technical Grade Phase Change Heat Storage Materials for Solar Heating Applications* (American Society of Mechanical Engineers, New York, NY (United States), 1995).
33. M.M. Farid, A.K. Mohamed, Effect of natural convection on the process of melting and solidification of paraffin wax. Chem. Eng. Commun. **57**, 297–316 (1987)
34. M.M. Farid, A. Kanzawa, Thermal performance of a heat storage module using PCM's with different melting temperatures: mathematical modeling. J. Sol. Energy Eng. **111**, 152–157 (1989)
35. M.M. Farid, R.M. Husian, An electrical storage heater using the phase-change method of heat storage. Energy Convers. Manag. **30**, 219–230 (1990)
36. Y. Chatani, H. Anraku, Y. Taki, Phase transition and structure change of urea adducts with n-paraffins and paraffin-type compounds. Mol. Cryst. Liq. Cryst. **48**, 219–231 (1978)
37. M.M. Farid, A.M. Khudhair, S.A.K. Razack, S. Al-Hallaj, A review on phase change energy storage: materials and applications. Energy Convers. Manag. **45**, 1597–1615 (2004)
38. A. Sahay, V.K. Sethi, A.C. Tiwari, M. Pandey, A review of solar photovoltaic panel cooling systems with special reference to Ground coupled central panel cooling system (GC-CPCS). Renew. Sustain. Energy Rev. **42**, 306–312 (2015). https://doi.org/10.1016/j.rser.2014.10.009
39. G. Alva, X. Huang, L. Liu, G. Fang, Synthesis and characterization of microencapsulated myristic acid–palmitic acid eutectic mixture as phase change material for thermal energy storage. Appl. Energy **203**, 677–685 (2017)
40. Y. Yuan, N. Zhang, W. Tao, X. Cao, Y. He, Fatty acids as phase change materials: a review. Renew. Sustain. Energy Rev. **29**, 482–498 (2014)
41. G.A. Lane, D.N. Glew, E.C. Clark, H.E. Rossow, S.W. Quigley, S.S. Drake, J.S. Best, Heat of fusion systems for solar energy storage, in: *Proceedings in Workshop on Solar Energy Storage Subsystems Heating and Cooling of Building* (1975), pp. 43–55
42. C.S. Herrick, D.C. Golibersuch, Qualitative Behavior of a New Latent Heat Storage Device for Solar Heating/Cooling Systems. General Electric Company Corporate Research and Development (1977)
43. A. Solé, H. Neumann, S. Niedermaier, I. Martorell, P. Schossig, L.F. Cabeza, Stability of sugar alcohols as PCM for thermal energy storage. Sol. Energy Mater. Sol. Cells. **126**, 125–134 (2014)
44. A. Sharma, V.V. Tyagi, C.R. Chen, D. Buddhi, Review on thermal energy storage with phase change materials and applications. Renew. Sustain. Energy Rev. **13**, 318–345 (2009)
45. J.H. Dieckmann, *Latent Heat Storage in Concrete* (University of Kaiserslautern, Germany, 2006).
46. H. Feilchenfeld, S. Sarig, Calcium chloride hexahydrate: a phase-changing material for energy storage. Ind. Eng. Chem. Prod. Res. Dev. **24**, 130–133 (1985)

47. K.C. Ting, P.N. Giannakakos, S.G. Gilbert, Durability of latent heat storage tube-sheets. Sol. Energy. **39**, 79–85 (1987)
48. F.C. Porisini, Salt hydrates used for latent heat storage: corrosion of metals and reliability of thermal performance. Sol. Energy **41,** 193–197 (1988)
49. B. Zalba, J.M. Marın, L.F. Cabeza, H. Mehling, Review on thermal energy storage with phase change: materials, heat transfer analysis and applications. Appl. Therm. Eng. **23**, 251–283 (2003)
50. S.D. Sharma, K. Sagara, Latent heat storage materials and systems: a review. Int. J. Green Energy. **2**, 1–56 (2005)
51. G.A. Lane, Phase change thermal storage materials, in *Handbook of Thermal Design*, ed. by C. Guyer (McGraw Hill B. Co, 1989)
52. S. Marks, An investigation of the thermal energy storage capacity of Glauber's salt with respect to thermal cycling. Sol. Energy **25**, 255–258 (1980)
53. H. Kimura, J. Kai, Phase change stability of CaCl2· 6H2O. Sol. Energy **33**, 557–563 (1984)
54. M. Liu, W. Saman, F. Bruno, Review on storage materials and thermal performance enhancement techniques for high temperature phase change thermal storage systems. Renew. Sustain. Energy Rev. **16**, 2118–2132 (2012)
55. A.I. Fernández, C. Barreneche, M. Belusko, M. Segarra, F. Bruno, L.F. Cabeza, Considerations for the use of metal alloys as phase change materials for high temperature applications. Sol. Energy Mater. Sol. Cells. **171**, 275–281 (2017)
56. C.E. Birchenall, M. Telkes, Thermal storage in metals. Shar. Sun Sol. Technol. Seventies **8**, 138–154 (1976)
57. A.M. Gasanaliev, B.Y. Gamataeva, Heat-accumulating properties of melts. Russ. Chem. Rev. **69**, 179 (2000)
58. D. Farkas, C.E. Birchenall, New eutectic alloys and their heats of transformation. Metall. Trans. A. **16**, 323–328 (1985)
59. C.E. Birchenall, A.F. Riechman, Heat storage in eutectic alloys. Metall. Trans. A. **11**, 1415–1420 (1980)
60. F. Li, Y.J. Hu, R.Y. Zhang, The influence of heating-cooling cycles on the thermal storage performances of Al-17 wt.% Si alloy. Adv. Mater. Res., 2248–2251 (2011)
61. J.P. KotzÃŠ, T.W. von BackstrÃ¶m, P.J. Erens, High temperature thermal energy storage utilizing metallic phase change materials and metallic heat transfer fluids. J. Sol. Energy Eng. **135**, 35001 (2013)
62. T. Akiyama, Storage and release of heat in a single spherical capsule containing phase change material of high melting point. Heat Transf. Japanese Res. **21**, 199–217 (1992)
63. X. Wang, J. Liu, Y. Zhang, H. Di, Y. Jiang, Experimental research on a kind of novel high temperature phase change storage heater. Energy Convers. Manag. **47**, 2211–2222 (2006)
64. J.Q. Sun, R.Y. Zhang, Z.P. Liu, G.H. Lu, Thermal reliability test of Al–34% Mg–6% Zn alloy as latent heat storage material and corrosion of metal with respect to thermal cycling. Energy Convers. Manag. **48**, 619–624 (2007)
65. E. Risueño, S. Doppiu, J. Rodríguez-Aseguinolaza, P. Blanco, A. Gil, M. Tello, A. Faik, B. D'Aguanno, Experimental investigation of Mg-Zn-Al metal alloys for latent heat storage application. J. Alloys Compd. **685**, 724–732 (2016)
66. R.J. Petri, E.T. Ong, M. Olszewski, *High Temperature Composite Thermal Storage Systems* (Institute of Gas Technology, Chicago, IL (USA); Oak Ridge National Lab, TN, 1984)
67. A.H. Abedin, M.A. Rosen, A critical review of thermochemical energy storage systems. Open Renew. Energy J. **4** (2011)
68. J. Yan, C.Y. Zhao, Experimental study of CaO/Ca (OH) 2 in a fixed-bed reactor for thermochemical heat storage. Appl. Energy **175**, 277–284 (2016)
69. D. Aydin, S.P. Casey, S. Riffat, The latest advancements on thermochemical heat storage systems. Renew. Sustain. Energy Rev. **41**, 356–367 (2015)
70. H. Zondag, B. Kikkert, S. Smeding, R. de Boer, M. Bakker, Prototype thermochemical heat storage with open reactor system. Appl. Energy **109**, 360–365 (2013)

71. A.H. Abedin, M.A. Rosen, Closed and open thermochemical energy storage: energy-and exergy-based comparisons. Energy **41**, 83–92 (2012)
72. N. Yu, R.Z. Wang, L.W. Wang, Sorption thermal storage for solar energy. Prog. Energy Combust. Sci. **39**, 489–514 (2013)

Chapter 3
Advanced Thermal Energy Storage Materials

3.1 Summary

The various advanced energy storage materials are discussed in this chapter. The advanced energy storage materials have better thermal characteristics compared to conventional energy storage and significant capacity for thermal energy storage. These include molten salts nanofluids, composite PCM that contains encapsulated PCM, nanoparticle-based nano PCM and cascade PCM, hybrid PCM and MXene-based PCM.

3.2 Various Advanced Thermal Energy Storage Materials

As the performance of conventional energy storage materials is very low due to poor storage capacity, low value of TC and high energy losses, so, various advanced energy storage materials are discussed in detail, which have better thermal properties and exhibit better properties compared to conventional energy storage materials.

3.2.1 Molten Salts and Molten Salt-Based Nanofluids

Molten salts are an important class of fluids like fluorides, nitrates, carbonates in which sensible or latent heat is stored [1]. The term molten salt refers to a liquid obtained from fusion of inorganic salt and has low vapor pressure, stable at high temperature, high heat capacity and low viscosity [2, 3]. Molten salts are considered better in performance compared to conventional energy storage as molten salts have high temperature range up to 600 °C [4]. Molten salt is also used as liquid storage medium when the temperature of heat transfer medium is above 400 °C and maximum operating temperature for molten salt is 565 °C in this case [5]. But molten salts

H. M. Ali et al., *Thermal Energy Storage*,
https://doi.org/10.1007/978-981-16-1131-5_3

having melting temperature 250 °C are suitable for working conditions. Hitec solar salt mixture is very popular having melting point 142 °C [6]. Molten salts are mostly used in concentrated solar power (CSP) applications.

Flueckiger et al. [7] suggested thermocline low-cost molten salt tanks for thermal energy storage. The key issues associated with design and performance were identified and indicated that molten salts flow rate decreased the axial extent of heat exchange and enhanced discharge efficiency. However, the economic assessment of thermocline tank and maintenance cost was needed. Herrmann et al. [8] reviewed the efficiency of molten salt vs steam in concentrated solar power plant and schematic diagram is shown in Fig. 3.1. The economic analysis was also conducted of TES system using levelized cost of electricity (LCOE) calculations. The direct molten salt TES was best for large capacities as compared to steam accumulator and lowest electricity prices achieved using LCOE calculations.

Devaradjane [9] suggested solar salt ($NaNO_3$-KNO_3) in concentrated solar power (CSP) system as thermal energy storage medium. The solar salt stored extra thermal energy during daytime, which was utilized at any time specially for high power demand. The major challenge was the freezing point of molten salt, which was 220 °C and it was reduced by adding calcium nitrate $Ca(NO_3)_2$ to about 120 °C. Muñoz-Sánchez et al. [10] summarized studies on molten salts base nanofluids of past work and also described the associated factors. To check the feasibility of molten salt, experimental and research studies were conducted. The concluding remarks were that still a lot of research needed in the field of molten salts.

Chieruzz et al. [11] developed various nanofluids by mixing molten salt-based fluids with nanoparticles using direct synthesis technique and thermal properties were investigated. The base salt was $NaNO_3$–KNO_3 as 60:40 ratio and nanoparticles were

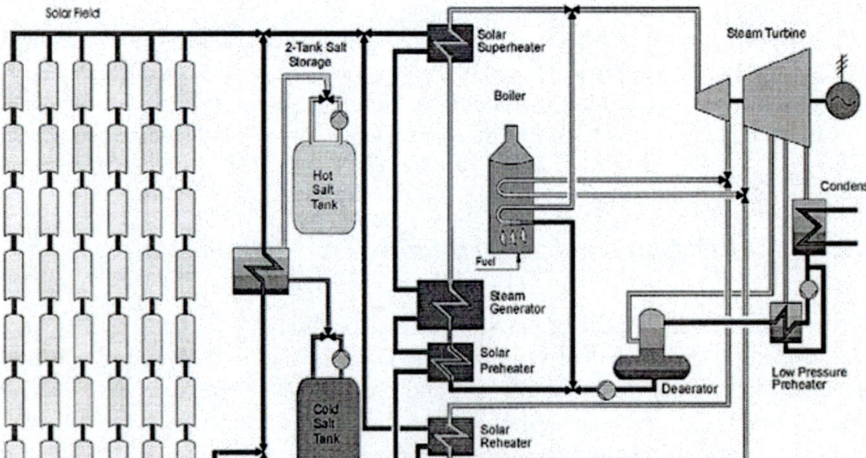

Fig 3.1 Schematic of parabolic trough tower along two-tank molten salt [8]

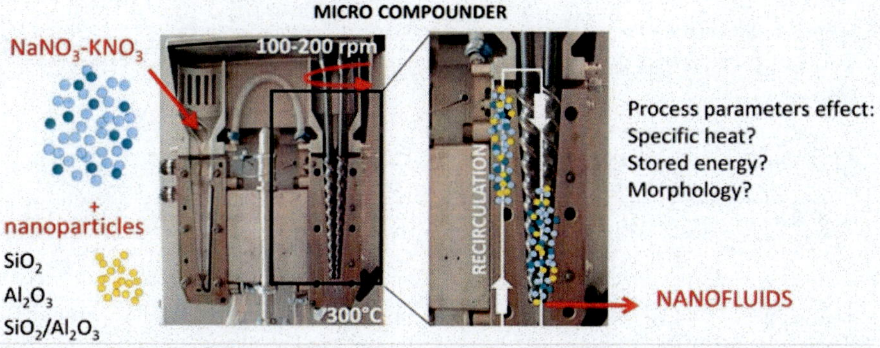

Fig 3.2 Twin screw micro-compounder for mixing nanoparticles with base fluid [12]

silica (SiO_2), alumina ($Al2O_3$), titanium dioxide (TiO_2) and SiO_2–Al_2O_3, which were mixed in water to form nanofluids. The results showed that specific heat increased from 15 to 57% and 1 to 22% in solid and liquid phase, respectively. Chieruzz et al. [12] also investigated molten salt-based hybrid nanofluids for thermal energy storage in concentrated solar power applications. The base fluid was $NaNO_3$–KNO_3 and nanoparticles were SiO_2, $Al2O_3$, TiO_2 and SiO_2-Al_2O_3 with concentration of 1.0 wt%. The apparatus diagram is illustrated in Fig. 3.2. The results indicated that heat capacity of base fluid in solid phase was enhanced with SiO_2-Al_2O_3 from 1.589 J/g °C to 2.417 J/g °C as compared to liquid phase, which was increased only from 1.632 J/g °C to 1.936 J/g °C.

Navarrete et al. [13] improved TES of solar salt (60% $NaNO_3$–40% KNO_3) by adding PCM-based Al–Cu alloy with nanoencapsulation of alumina (Al_2O_3) layer naturally formed when exposed to air. The specific heat and TES were evaluated at various mass loads. The total thermal energy storage was increased up to 17.8% at constant volume base comparison. The summary of various authors investigated molten salts nanofluids is given in Table 3.1.

3.2.2 *Composite PCMs*

The thermal storage capacity of PCMs can be enhanced by encapsulation of PCMs, adding NPs having high value of TC, which enhances the thermal characteristics of PCMs, and the use of multiple PCM called cascade PCMs. Several studies have been done by various researchers on composite PCMs, which are discussed in detail in the below sections.

Table 3.1 Summary of experimental work of molten salt-based nanofluids

Research institution	Base fluid	Melting temperature (°C)	NPs	References
ENEA Italy	Solar salt (50 mol $NaNO_3$-50 mol KNO_3)	222	Al_2O_3	[11, 14]
ENEA Italy	KNO_3	334	Al_2O_3, SiO_2	[11, 14]
Texa A & M university	$BaCl_2$-NaCl-$CaCl_2$-LiCl (15.9:34.5:29.1:20.5 mol)	378	SiO_2	[15]
University of Texas	Li_2O_3-K_2CO_3 (62:38 mol)	488	MgO	[16]
MIT	$NaNO_3$	308	Al_2O_3, SiO_2, TiO_2	[17]
National Tsing Hua University, Taiwan	Hitec salt KNO_3-$NaNO_3$-$NaNO_2$ (53:40:7 mol)	222	Al_2O_3	[18]
North china Electric Power University, China	Solar salt (50 mol $NaNO_3$-50 mol KNO_3)	222	MWCNTs	[19]
University of Leeds, UK	Solar salt (50 mol $NaNO_3$-50 mol KNO_3)	222	CuO	[20]
Energy development center, Chile	Solar salt	222	Al_2O_3, SiO_2	[21]
–	Molten nitrate salt	250–400	SiO_2@ Al_2O_3	[22]
Beijing University, China	Quaternary mixed nitrates of calcium, potassium, sodium, lithium	–	SiO_2	[23]
–	$NaNO_3$	308	SiO_2	[24]
CIC Albert Einstein, Spain	Eutectic carbonate salt	–	Al_2O_3	[25]
–	Solar Salt (60 wt% $NaNO_3$-40 wt% KNO_3)	222	Al_2O_3	[26]
South China University	Solar salt	222	MgO	[27]

3.2.2.1 Encapsulation of PCM

The energy storage capacity can be enhanced by utilizing encapsulated phase change materials. In this, PCM is enveloped in solid structure called shell, which is made of plastic or synthetic polymers [28]. It may be micro or nano encapsulation. The encapsulated PCM has two parts, namely, core of PCMs and shell of polymer of inorganic material. The shell material may be core shell, multi shell or polynuclear structure [29] as represented in Fig. 3.3.

Table 3.2 Summary of work done by various authors on metallic NPs addition in energy storage materials

References	Nanoparticle	PCM	Remarks
[62]	Al_2O_3	Paraffin Wax	The latent heat enhanced by 10.8% by adding Al_2O_3 of 10 nm sized but TC was low
[63]	Al_2O_3	Paraffin Wax	The addition of 10–20 nm Al_2O_3 nanoparticles boosted 1–43% TC. Where carbon black nanoparticles having 15–20 nm size enhanced the thermal energy storage by 35%
[64]	TiO_2	Palmitic acid	TiO_2 nanoparticles of sized 21 nm enhanced TC of PCM by 80% and stability of palmitic acid was also improved
[65]	Al_2O_3	Paraffin Wax	The addition of nanoparticles in base PCM enhanced viscosity, melting process and TC but the value of "h" decreased by increment in viscosity
[66]	CuO	n-eicosane	The melting rate of nano-enhanced PCM increased with the addition of nanoparticles from 0% to 5% and base temperature of heat sink lowered by 4 °C
[67]	CuO	Cyclohexane	Nano-enhanced PCM melted fast at higher volume fraction of nanoparticles and less energy needed for this process
[68]	CuO	–	The solidification and freezing time decreased by adding nanoparticles in PCM, but energy storage capacity was also reduced
[69]	CuO	–	The melting rate was enhanced by adding nanoparticles but decreased with increasing volume fraction due to viscosity enhanced

There are various methods for the encapsulation of PCM namely mechanical, chemical or physical methods and procedures have been explained by different authors [30, 31]. The physical methods include spray drying and fluidized bed process. The chemical process includes interfacial polymerization, coacervation, phase separation, in-situ-polymerization [32]. The physic-chemical techniques included ionic gelation, coacervation, sol–gel method. The different methods are used for microencapsulation [33], which includes pan coating, air-suspension coating, centrifugal extrusion, vibrational nozzle and solvent evaporation. The encapsulation ratio (R) [34] can be defined as

$$R = \frac{H_{M,MicroPCM}}{H_{M,PCM}} \tag{3.1}$$

where $H_{M,MicroPCM}$ and $H_{M,PCM}$ are the heat of fusion of microencapsulated PCM and only PCM. Microcapsules have shapes like tabular, spherical or oval as illustrated in Fig. 3.4.

Table 3.3 Various studies on nanostructures composite PCMs by different authors

Base fluid	Composite material	Concentration (wt%)	Applications	Thermal conductivity enhancement	References
Palmitic acid	CNFs	0.2–5 wt%	Thermal energy storage	The value of "k" of M-CNF/PA with 0.2, 0.5% wt of M-CNF higher than 2.0, 5.0% wt of M-CNF	[104]
Palmitic acid	PDMS-G-NF with Ni-foam	–	Thermal energy storage	The latent heat was 42.3–123.41 kJ/kg of composite and value of TC was 2.262 W/mK at 59.02% wt of PCM composite	[101]
Paraffin	CNTs	–	Thermal energy storage	The value of TC of composite was 1.2 W/mK at 80 wt% of paraffin wax	[85]
Soy wax, Paraffin	CNTs, MWCNTs	1,2,5,10 wt%	Thermal energy storage	The "k" value of pure paraffin is 0.320 W/mK and "k" value of PCM composite is 0.450 W/mK having 10 wt% of CNT	[78]
Paraffin wax	Functionalized-MWCNTs	–	Thermal energy storage	The value of "k" of about 86.7% for f-MWCNT as compared to pure paraffin wax	[81]
Paraffin wax	MWCNTs	–	Thermal energy storage	The results showed that 75% weight of paraffin RT-50 and MWCNT have the value of "k" about 0.58 W/mK	[82]
Steric acid	MWCNTs/graphene/graphite	–	Thermal energy storage	The addition of MWCNTs, graphene, graphite enhanced TC 73%, 35%, 188% respectively	[105]

(continued)

Table 3.3 (continued)

Base fluid	Composite material	Concentration (wt%)	Applications	Thermal conductivity enhancement	References
Paraffin wax	MWCNTs	–	Thermal energy storage	Increment of TC in carbon foam owing to decrement of 11.5% and 7.8% surface temperature having porosity lower than 75% and 88% respectively	[106]
Lauric acid	GNP, MWCNTs	1.0 vol%	Thermal energy storage	The maximum TC enhancement was 223%, 171%, 27% for GNPs, MWCNTs, SWCNTs respectively	[107]
Beeswax	Graphene nanoparticles	–	Thermal energy storage	The latent heat and TC increased up to 31.98% and 1056% by addition of nanoparticles, respectively. The melting point reduced to 15.43 °C and thermal energy storage boost to 12%	[108]
Lauric acid	Graphene nanoparticles	–	Thermal energy storage	The graphene nanoparticles having thickness 5–10 nm boost TC 230% and thermal interface conductance was 100 MW/m^2K b/w lauric acid and graphene	[109]
Paraffin wax	Nano graphite	–	Thermal energy storage	The addition of nano graphite NPs enhanced the TC of PCM about 741%	[110]

Table 3.4 Various hybrid PCMs studied by different authors

Base fluid	Composite Material	Concentration (wt%)	Applications	Thermal Conductivity Enhancement	References
Expanded perlite	CNTs	–	Thermal energy storage	The TC was enhanced 113% by addition of CNTs, storage and release time decreased by 57.4% and 78.6% respectively	[135]
PEG	GO/BN/PEG	4 and 30	Thermal energy storage	The value of TC was enhanced by 900% compared to pure PEG	[137]
PEG/SiO_2	Carbon fiber/PCM	1–5	Thermal energy storage	The TC was 0.45 W/mK and 3 wt% gave the best results	[138]
PEG	GO/GNPs	–	Solar energy storage	The TC increment was 1.43 W/mK from 0.31 W/mK of pure PEG	[139]
Adipic acid/Sebacic acid	GNPs/PCM	0.1,0.3.0.5	–	The value of "k" was increased by 6.3%, 10.9% and 12.7% at 0.1 wt%, 0.3 wt% and 0.5wt% respectively	[140]
Polyethylene glycol/Diatomite	Ethylene glycol/PCM	10	–	The value of "k" of hybrid PCM was enhanced by 103%	[141]

(continued)

Table 3.4 (continued)

Base fluid	Composite Material	Concentration (wt%)	Applications	Thermal Conductivity Enhancement	References
Capric acid/Palmitic acid/Diatomite	Ethylene glycol/PCM	3,5	Thermal energy storage	The value of "k" enhanced by 53.7% in liquid state of hybrid PCM at 5 wt% of ethylene gylcol	[142]

Fig 3.3 Different structures of shell material [29]

Adeel et al. [35] reviewed the various techniques for the encapsulation of PCMs along which method was suitable for which type of PCM material. The preparation methods, thermal and morphological characteristics and stability were discussed extensively. The authors suggested that in-situ polymerization was the best method for preparation due to high encapsulation efficiency and less leakage. These encapsulated materials were especially useful for thermal energy storage applications. The inorganic PCMs are considered to be the best due to high value of TC but their encapsulation efficiency and unstability are still low, which limits its applications. Milian et al. [36] also reviewed the preparation techniques of encapsulated inorganic PCMs. The authors suggested two types of encapsulated inorganic PCMs, namely, core shell PCM and shape stabilized PCMs. For the first, inverse Pickering emulsion, electroplating, solvent evaporation precipitation and mechanical packaging methods were suggested for the preparation while for second, sol–gel, infiltration and impregnation encapsulation methods were suggested.

Many studies focused on encapsulation with various types of PCMs with shell. Zhang et al. [37] investigated the encapsulation of n-octadecane and concluded that during melting, the temperature of upper region was higher than lower region, which enhanced the melting process. This occurred due to natural convection of liquid phase, which was created as hot fluid temperature increased and cold fluid temperature decreased. The surface temperature increased from top to bottom as melting

Fig 3.4 The various morphologies of microcapsule [33]

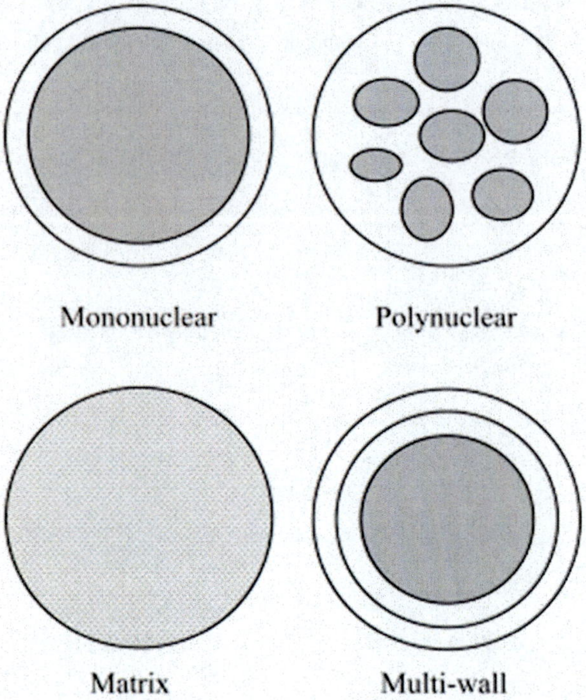

proceeded. Regin et al. [38] investigated the melting behavior of paraffin wax, which was in capsule with hot water around it. The experiments were performed, and results illustrated that melting was high in top region as compared to lower region and numerical as well as experimental results can only be agreed after considering natural convection in numerical modeling. Yuan et al. [39] synthesized microencapsulated PCM by immersing paraffin in silica. Graphene oxide (GO) NPs were dispersed in paraffin/SiO_2 to increase the TC. The value of TC for paraffin was enhanced by 25.82% by using this combination and encapsulation efficiency reached 49.6%. The TC of paraffin, paraffin/SiO_2, paraffin/SiO_2/GO was 0.3, 1.031 and 1.162 W/mK, respectively. Yu et al. [40] fabricated microencapsulated PCM using n-octadecane as core and calcium carbonates as shell by self-assembly technique. When the mass ratio of n-octadecane/calcium carbonate (CaCO3) was 30/70 then the TC was significantly increased by 9.1% compared to base fluid and SEM image of n-octadecane/CaCO3 is illustrated in Fig. 3.5.

Chai et al. [41] made microencapsulated PCM by n-eicosane into titanium dioxide through in-situ polycondensation in sol–gel process. The results showed that the ratio of composite with weight ratio 40/60 has TC of 0.865 W/mK and it was 18.6% high than pure n-eicosane. The TC enhancement was five times higher when compared to pure PCM.

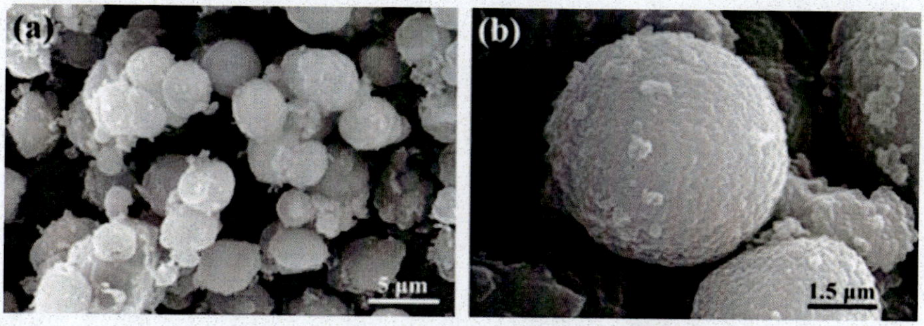

Fig 3.5 SEM image of microcapsule n-octadecane/CaCO3 with mass ratio 30/70 [40]

Safdari et al. [42] carried out a numerical study on encapsulated PCM in thermal management of battery system for active as well as passive system. The PCM surrounded by battery cell was encapsulated in circular, hexagonal and rectangular vessels. The results revealed that the thermal behavior of hexagonal and circular PCM was the same while efficient cooling was obtained by rectangular shape encapsulated PCM. Parveen and Suresh [43] investigated thermal performance of microencapsulated paraffin PCM in heat sink applications, which was prepared by in-situ polymerization. The experiments were conducted with fins as well as without fins and thermal stability along degradation of PCM was under-considered parameters. The results revealed that uniform temperature distribution and complete melting were occurred, and recovery time of heat sink was also decreased radically. Seitz and Ajiro [44] prepared encapsulated ultra-thin PCM by using icosane coating on acrylic acid and results revealed high storage capacity and good thermal stability. There was a minor alteration in reliability and repeatability of encapsulated PCM even after 100 cycle experiments. Zhang et al. [45] prepared first ester that was then encapsulated as core material by emulsion polymerization and then it was characterized to determine various aspects. The PCM showed high storage capacity having latent energy of 118 J/g and along with perfect spherical shape with excellent thermostability. The microencapsulation of paraffin wax with polyurethane was done by Nikpourian et al. [46] using interfacial polymerization. The effect of core/shell mass ratio on encapsulation efficiency was determined, and results concluded that energy storage capacity as well as energy storage efficiency of prepared material were 97.5% and 80.2%, respectively. The thermal resistance of encapsulated PCM was good after 100 cooling/heating cycles. Ho et al. [47] investigated microencapsulated PCM (MEPCM) at back side of panel for improvement in cooling effects. The MEPCM was placed on water surface represented in Fig. 3.6. The different parameters like air temperature, solar radiations, wind speed, water flow rate and temperature were under consideration. The electrical efficiency of system was improved by 2.1 % with 5 mm thickness and 30 °C melting temperature of MEPCM compared to untreated PV panel.

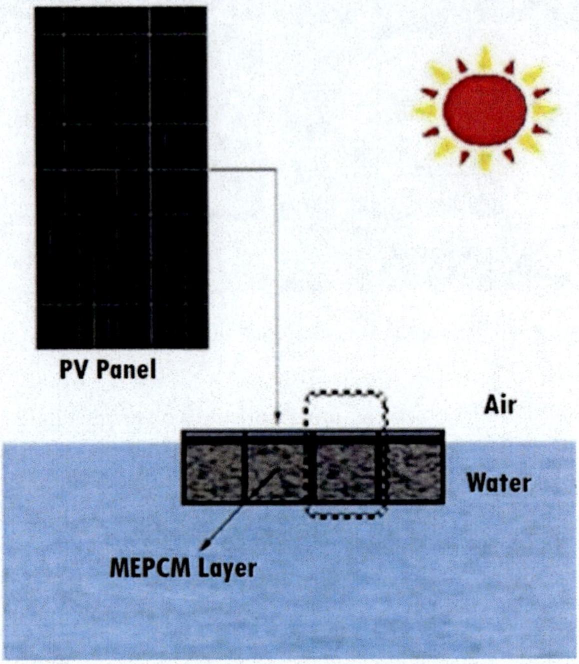

Fig 3.6 The MEPCM-PV system floating on water surface [47]

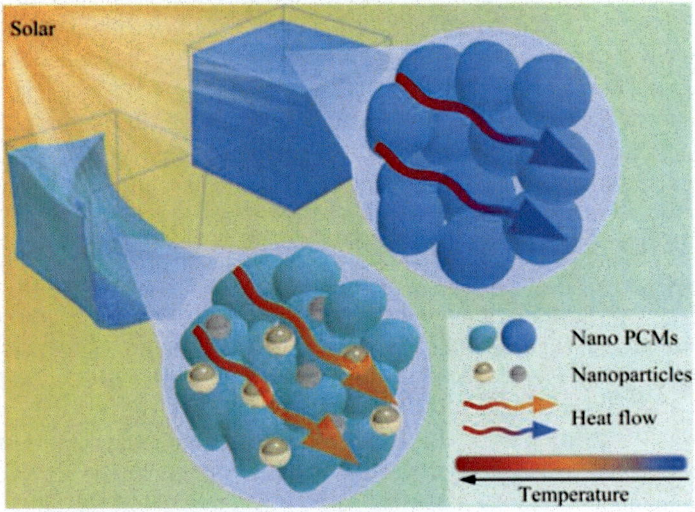

Fig 3.7 Addition of high TC nanoparticles in PCM for increment in thermal energy storage [48]

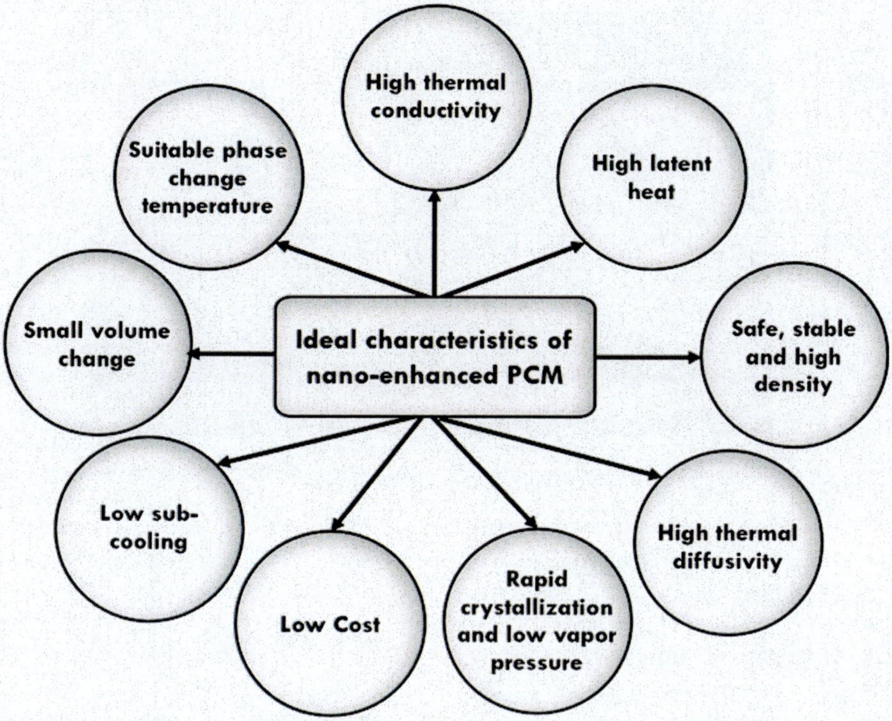

Fig 3.8 Ideal characteristics of nano-enhanced PCM [50]

Overall, the method of encapsulation of PCM reduces the leakage, subcooling and segregation issues but decrement of melting temperature and latent heat is the drawback of encapsulated PCM.

3.2.2.2 Addition of Nanoparticles in PCM

The PCMs have high thermal energy storage capacity with broad range of temperature but their low thermal conductivity limits the applications due to low release speed. Thus, the addition of high conductive materials enhances [48] the release speed of PCM via boosting the value of "k". The thermal increment of PCM may be done by nanoparticle addition, which greatly depends on physical properties and filling ratio.

Khodadadi and Hosseinizadeh [49] were the first researchers that investigated the enhancement in performance of PCM by dispersion of nanoparticles. They conducted computational study on nanoparticles enhanced PCM in thermal storage system. The copper nanoparticles and water were used, and the study revealed that as nanoparticle volume fraction increased the solidification time decreased resulted in an increment in TC compared to base fluid. The ideal characteristics of nano-enhanced PCM are:

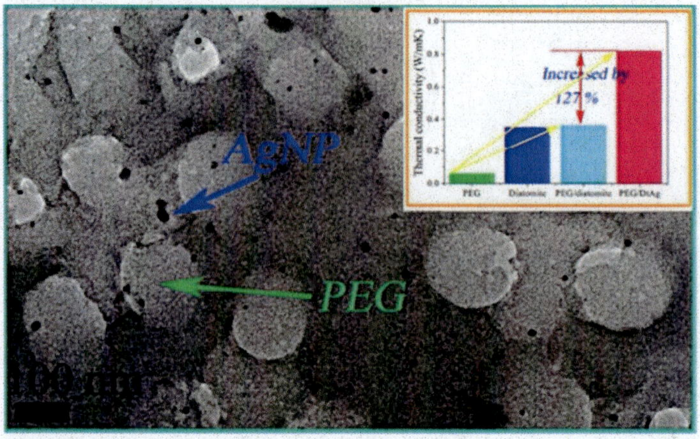

Fig 3.9 Enhancement in TC of AgNPs in PEG/diatomite [52]

- High thermal conductivity
- High latent heat
- Safe, stable and high density
- High thermal diffusivity
- Low vapor pressure
- Low subcooling
- Small volume change
- Suitable phase change temperature.

Metallic-Based Nanoparticles in PCM

The addition of metal nanoparticles in PCM is a traditional way to enhance the TC of PCM. The metal particles include copper, gold, titanium, silica and aluminum. Khodadadi and Fan [51] conducted numerical analysis on freezing of single PCM and nanoparticle-based composite PCM. The PCM was cyclohexane while nanoparticles were alumina, copper, copper oxide and titanium oxide. The results indicated that freezing time decrement was 11.36% due to suspension of nanoparticles and TC of composite PCM was increased.

Qian et al. [52] investigated gold nanoparticles (AgNPs) for thermal energy storage in polyethylene glycol (PEG)/diatomite and compared performance with PEG/diatomite. AgNPs having a diameter range 3–10 nm were used. The SEM image of AgNPs in PEG/diatomite is shown in Fig. 3.9. The results indicated that TC of PEG/diatomite/Ag was 0.82 W/mK greater about 127% compared to PEG/diatomite. The storage and released time were in decrement by 34 and 25% than PEG/diatomite.

Wu et al. [53] developed a new sort of material by suspending copper nanoparticles in melting paraffin. The copper-based nano PCM showed good stability and differential scanning calorimetry showed lower latent heat compared to pure paraffin. The

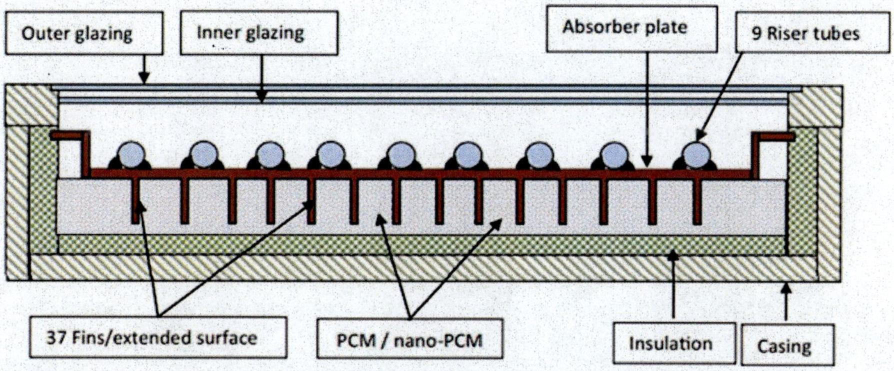

Fig 3.10 Schematic diagram of solar-TES integrated collector [55]

cooling and heating rates were significantly improved by adding copper nanoparticles in paraffin. The heating and cooling time of composite having 1 wt% Cu nanoparticles is reduced by 30.3% and 28.2%, respectively. Zeng et al. [54] investigated increment in TC of organic PCM by addition of gold nanoparticles. The composite consisted of 1-tetradecanol/Ag nanoparticles, which contained 62.73 wt% of gold nanowires. The value of TC of composite was about 1.46 W/mK and high phase change enthalpy of 76.5 J/kg. Gold nanoparticles were strong candidates for increment in TC of organic PCM.

Kayiem and Lim [55] conducted experiments on flat plate collector integrated with thermal energy storage. The paraffin wax and composite of paraffin wax containing 1 wt% of 20 nm copper nanoparticles were used at different inclination angles of collector and fins were also used as represented in Fig. 3.10. The measurement results showed that by using Cu nanoparticles, efficiency enhanced up to 21.7% compared to conventional system.

The metal oxide NPs addition in PCM also enhances storage capacity due to having a large value of TC. The metal oxides are stable after reaction of metals with oxygen and similar is the case with non-metal oxides. Attia et al. [56, 57] investigated Al_2O_3 nanoparticles to enhance the properties of thermal storage system. The NPs were mixed and aqua ethylene glycol solution of 50 wt% as heat transfer fluids at various volume concentrations was also used. The NP addition to PCM had a better advantage on cooling process and a combination of these techniques played vital effect on the reduction in charging time. The percentage reduction of charging time at heat transfer fluid inlet temperature of −12 °C was 32%, 28%, 18%, 12% at flow rate of 12, 10, 8, 6 L/min respectively.

Elsayed [58] used various nanoparticles for thermal energy storage with neopentyl-glycol as PCM. The nanoparticle included copper, aluminum, silica, titania (Cu, Al, SiO_2, TiO_2) and binary NPs. The binary nanoparticles containing (6% SiO_2 + 6% Al) showed 34.6% more storage than composite containing (6% Cu + 6% Al). The better results were indicated by composition (6% Al + 3% SiO_2 + 3% TiO_2) with neopentyl-glycol.

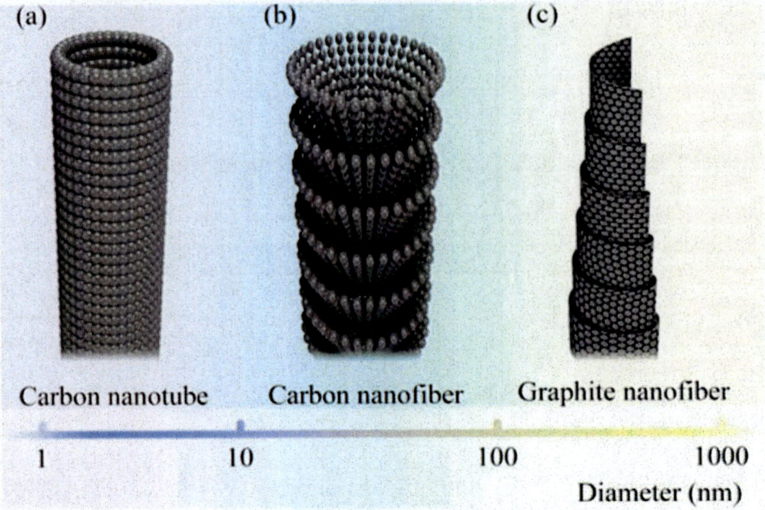

Fig 3.11 Structure for **a** CNTs **b** CNF **c** Graphite NF [48]

The study of Darzi et al. [59], in 2006, was done to improve melting and freezing rates of PCM by addition of nanoparticles, fins and changing inner tube shape. The 2% and 4% Cu nanoparticle addition in n-eicosane decreased melting time 25% and 46% and solidification time reduced to 9% and 16%, respectively. Mehdi and Nsofor [60] used nanoparticles, which were dispersed in PCM in triple tube heat exchanger and 17% melting time was saved by adding 1% volume fraction of nanoparticles. Babapoor et al. [61] investigated polyethylene glycol (PEG), polyamid6 and various NPs included (SiO_2, Al_2O_3, Fe_2O_3, ZnO) to form nano-fibers nanocomposite PCM prepared by the method of electrospinning. It was observed that fiber diameter decreased with increment in NPs loading and lowest average diameter was for Fe_2O_3. The Al_2O_3 showed the highest TC and nanocomposite exhibited high heat enthalpy for thermal energy storage applications.

Non-metallic-Based Nanoparticles in PCM

The non-metallic NPs include carbon-based NPs having graphene, graphite, carbon black and carbon nanotubes that occur in nanostructure. The nanostructures are low dimensional materials in which scale of length is reduced to few nanometers. The applications of nanostructures in nanolevel technology have been promising fields of material science research. The nanostructures include nanoparticles, metals in which crystals are distributed at nanoscales, capsules and materials having nanoscales holes [70].

Addition of Carbon Nanotubes

Carbon nanotubes are carbon allotropes having cylindrical nanostructures with diameter about one nanometer. The basic types of carbon nanotubes are [71, 72]:

- Single-Wall Carbon Nanotubes (SWCNTs)
- Multi-Wall Carbon Nanotubes (MWCNTs).

The single-wall carbon nanotubes are basic structures while multi-wall carbon nanotubes are coaxial cylinders having interlayer spacing. The name MWCNTs are referred to nanostructures having outer diameter less than 15 nm and the diameter above this are referred as carbon nanofibers [73]. CNTs are differentiated from nanofibers due to layered graphite sheets in nanofibers. By assuming the needle-like cylindrical shape of NPs, Yu and Choi [74] suggested the Sauter mean diameter (d) of carbon nanotube (CNT) nanoparticles represented as [75]

$$d_{CNT} = \frac{1.5a.b}{a + b/2} \approx 1.5b \tag{3.2}$$

where a is the length and b is the diameter of NPs.

Asseal et al. [76] used (CTAB) in their study as surfactant with MWCNT and compared the value of "k" in double-wall and multi-wall carbon nanotubes. They determined that at same vol%, the increment in "k" was less in double-wall compared to multi-wall CNTs and maximum enhancement was 7.6% at 1 vol% of double wall CNTs in water. Carbon has high TC and has five times higher TC than silver [77]. Cui et al. [78] used (MWCNT) as well as carbon nanofibers in paraffin to enhance the value of "k" of base PCM. The composite was prepared by dispersing NPs in paraffin at 60 °C and at 1, 2, 5, 10 wt% of NPs concentration. The "k" value of pure paraffin wax was 0.320 W/mK while for composite PCM, it was 0.450 W/mK having 10 wt% of MWCNT.

Wang et al. [79] mixed MWCNTs with base PCM to increase value of "k" of composite PCM, which was prepared by drafting method with hydroxyl group using mechochemical reaction. As palmitic acid (PA) mixed with MWCNTs, and composite doped in OH group, the value of "k" enhanced 51.6% with MWCNT of 1 wt%. Wang et al. [80] used (MWCNTs) NPs for increment in value of "k" of PCM to about 46%. The material was prepared by ball milling method having multi-wall CNTs in palmitic acid with KOH to increase dispersion. The MWCNT-PA composite having 1 wt% of MWCNT enhanced the value of "k" by 46% in solid state. Tang et al. [81] doped functionalized multi-wall carbon nanotubes (f-MWCNT) along n-octadecyl-amine in paraffin wax. The results indicated that increment in value of "k" of composite was about 86.7% for f-MWCNT as compared to pure paraffin wax. Gilart et al. [82] proposed PCM/carbon-based composite material that enhanced the value of "k" of composite by 576%. The MWCNT of NC700 having surface area of 271 m^2 was selected and it was not expanded. The developed composite was stable without any leakage of paraffin during phase change cycle and composite showed

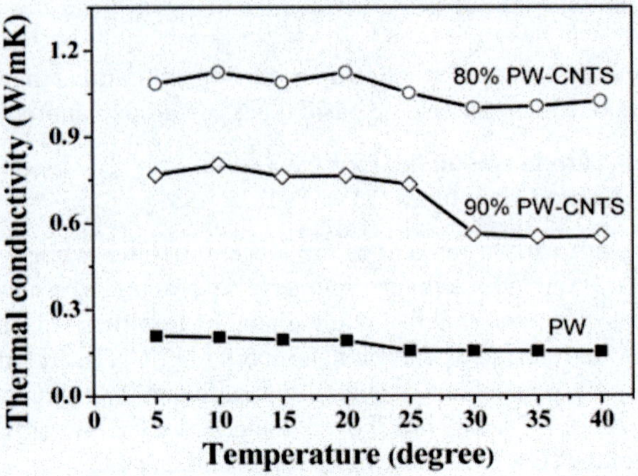

Fig 3.12 The value of "k" for paraffin and along 10% and 20% wt of CNT foam [85]

multifunctional characteristics. The results indicated that 75% weight of paraffin RT-50 and MWCNT have the value of "k" of about 0.58 W/mK. Vilatela et al. [83] investigated CNT composite and multifunctional composite, which were prepared by chemical vapor deposition and direct polymer infiltration method. The electrical and TC of CNT/epoxy composite demonstrated super functionally and composite consisted of epoxy matrix with high volume fraction. The results indicated that CNT fiber-enhanced thermal conductivity 157 W/mK per unit fiber mass fraction in base PCM.

Xing et al. [84] investigated the increment value of "k" by using short single, long single and multi-walled carbon nanotubes and compared with pure water. At concentration 0.48 vol% of CNTs at 60 °C, the CNTs showed maximum increment in value of "k" of 8.1, 16.2 and 5.0% for single, long and multi-walled carbon nanotubes and enhancement as linear with increasing concentration of CNTs. Also, the increment was maximum in case of long single-walled carbon nanotubes. Chen et al. [85] investigated carbon nanotube (CNT) foam for increment in value of "k" by 500% due to enhancement in latent heat storage capacity. The composite PCM was prepared by paraffin wax (PW) filled within flexible CNT-based porous material. As the value of "k" of CNTs is 3000 W/mK so by adding CNT the composite achieved value of "k" to approximate 1.2 W/mK at 80 wt% of (PW) resulted in six times increment effect represented in Fig. 3.12.

Assael et al. [86] used CNT nanofluids for increment in value of "k", which were prepared by catalytic decomposition method with outer and inner diameter of nanotubes 15–330 nm and 2–30 nm, respectively. The results indicated that 38% enhancement in value of "k" at 0.6 vol% of CNT/water and reduction in value of "k" for decrement in length of nanotube. Liu et al. [87] determined the value of "k" of MWCNTs in ethylene glycol and synthetic oil. At 1 vol% of CNT and ethylene glycol, the linear trend was observed in increment of "k" and 12.4% was observed at this concentration. While for synthetic engine oil at 2 vol%, the enhancement was

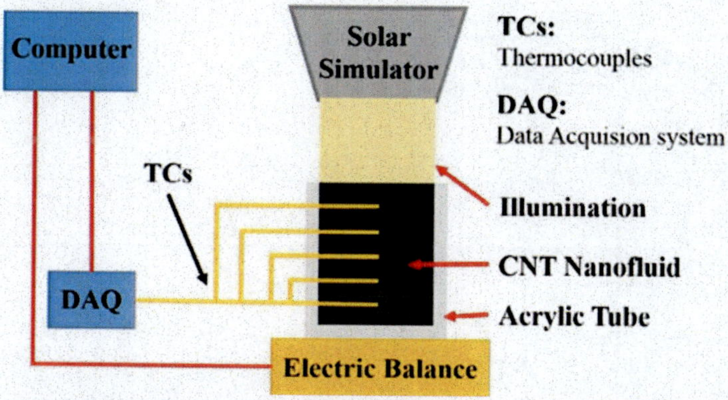

Fig 3.13 Steam generation by CNTs nanofluids [89]

30%. The authors also reported that CNT–oil nanofluids have 3–5 times higher value of "k" compared to the existing models.

Ding et al. [88] measured the value of "k" of MWCNTs nanofluids, which were prepared by sonicating CNT sample in ultrasonic bath for 24 h, dispersing in Gum Arabic and treat mixture with homogenizer for 30 min. The value of "k" largely depends on temperature and CNT concentration and mainly dependent on temperature. The results indicated that at 1 vol% and 30 °C, the increment in value of "k" was 80%.

Wang et al. [89] proposed model for solar steam generation by carbon nanotubes nanofluids and experiments were performed for this purpose as setup is illustrated in Fig. 3.13. The effect of CNTs concentration on solar steam generation was under-examined. The results indicated that high evaporation efficiency of 46.8% was observed with 19.04×10^{-4} vol% of CNTs under solar radiation of 10 KW/m^2.

Hashemi and Noie [90] investigated flow boiling heat transfer characteristics using (MWCNTs)/pure water nanofluids in horizontal tube. The nanofluids were prepared by using 1:1 ratio of MWCNTs and gum Arabic as surfactant. The boiling process of this nanofluid and pure water was observed specially after critical heat flux. The experimental results exhibited that the value of "h" was higher for nanofluids compared to pure water and increment was 4.3% at critical heat flux for MWCNTs-GA/water nanofluids as shown in Fig. 3.14.

Mahbubal et al. [91] studied the effect of single-wall CNTs (SWCNTs) nanofluids on enhancement in efficiency of evacuated tube solar collector. The thermal efficiency of collector working was compared by using water and that of nanofluids. The experimental data showed that increment of thermal efficiency of 56.7 and 66% was observed when water and 0.2 vol% nanofluid was used in collector. The critical solar irradiance of 900 W/m^2 was found and there was no increment in efficiency beyond this limit.

The nanotubes and NPs addition into base fluid increase TC of composite PCM. The TC of composite PCM [92] having MWCNTs/graphene boosted TC to 115.4%,

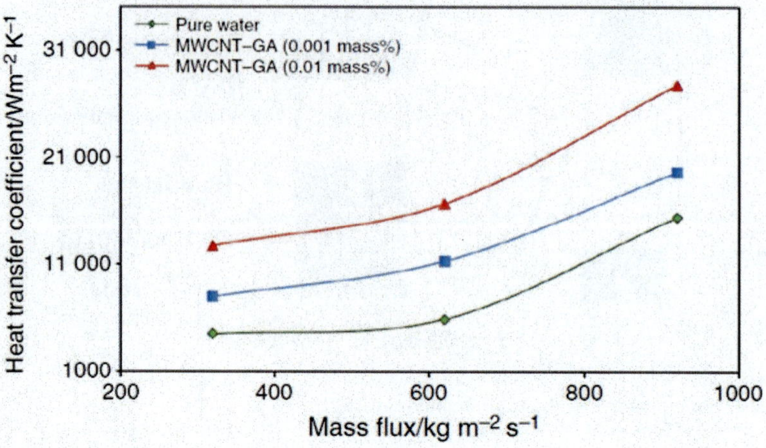

Fig 3.14 Experimental "h" for water and MWCNT-GA/water at critical heat flux [90]

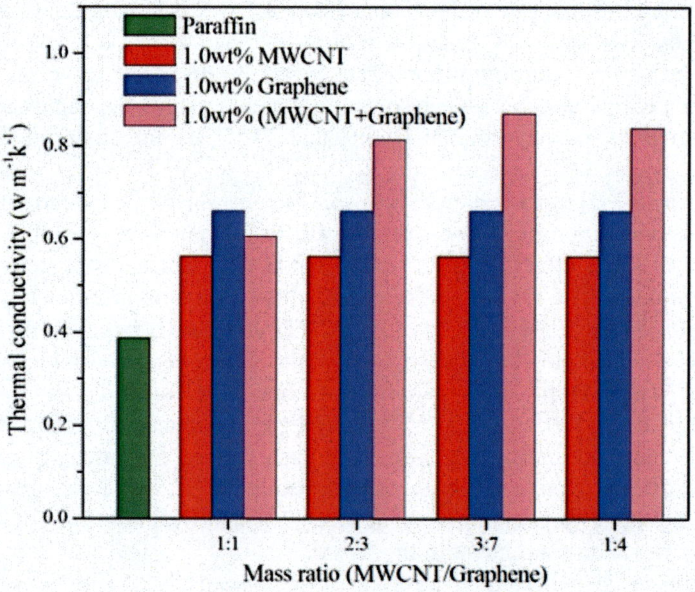

Fig 3.15 The TC increment of paraffin, MWCNTs, graphene, MWCNTs and graphene [92]

31.8%, 55.4% when compared with graphene-based composite PCM,MWCNT-based composite PCM and pure PCM, respectively, as represented in Fig. 3.15. The composite PCMs have high increment/decrement rate compared to pure paraffin.

Other Carbon-Based NPs Addition in PCM

Carbon nanofibers (CNFs) are like carbon nanotubes (CNTs) but they can be seen along the length of filament in planner layer which is different from CNTs. The unique structure of carbon nanofibers exhibits better semiconductor behavior and they are chemically active on inner and outer surfaces [93]. Carbon fibers take shape when mixed with PCM by melting-dispersion technique or hot press method as there is no microstructure of carbon fiber. The PCM/carbon fiber was prepared by Namura et al. [94] using hot-press method and laser flash technique but first was the best method. The value of "k" of composite with 25 vol% carbon fiber was 30 W/mK while it was 0.733 W/mK for pure PCM.

Cui et al. [78] dispersed CNFs in molten soy wax using ultrasonication to make composite PCMs. The TC of PCM enhanced up to 44.8%, which was also higher than CNTs. Yu et al. [95] studied the TC of paraffin wax, which was enhanced by 13.3% by addition of CNFs. Ismar and Sarac [96] investigated composite nanofibers containing lauric acid and polyacrylonitrile combined with PCMs, which were prepared by electrospinning technique. The composite formed was used as thermal energy storage in the patch form and can be used in heat treatment applications to enhance the total enthalpy of patch. Lu et al. [97] fabricated lauric acid (LA) containing polystyrene NFs, which has storage capacity of up to 78.4%. The X-ray diffraction pattern indicated that crystal size of LA domain became large along increment its loading in the composite. The supercooling effects decreased with percentage enhancement of LA in composite. The composite showed excellent stability in 100 cooling-heating cycles during temperature range of 0–80 °C.

Said et al. [98] investigated nano-enhanced PCM having calcium chloride hexhydrate with CNF for thermal energy storage. The power density of PCM was low, which resulted in decrement in the value of "h". However, the value was enhanced after the addition of carbon nanofibers having 1 wt% in PCM. The mathematical model was also modified based on the literature of geometry and results indicated that nano-enhanced PCM has high applications for thermal energy storage.

Kant et al. [99] investigated heat transfer behavior on melting of PCM with graphene nanoparticles in square cavity. The PCM used was capric acid, calcium chloride ($CaCl_2.6H_2O$) and n-octadecane with 1, 3, 5% volume ratio of graphene nanoparticles. The simulation results revealed that TC enhanced by the addition of nanoparticles but increment in viscosity resulted in decrement in the value of "k". Similar results were found by Sushoban and Kar [100] when thermal modeling was done on the melting of nano-enhanced PCM. CuO nanoparticles with high volume fraction increased the viscosity but reduced the value of "h". Liang et al. [101] investigated super-oleophilic graphene with porous nickel foam to increase TC of PCM. The composite was prepared by inserting layers of graphene on surface of nickel foam to form graphene nickel foam, which was further modified with polyydimethysiloxane to form shape stable PCM. The "k" value of this PCM was 2.262 W/mK at 59.02 wt% of PCM compared to only palmitic acid (PA).

Warzoha et al. [102] studied herringbone-style graphene NPs penetrated in paraffin wax. The TC of composite PCM enhanced in solid state by 80% while

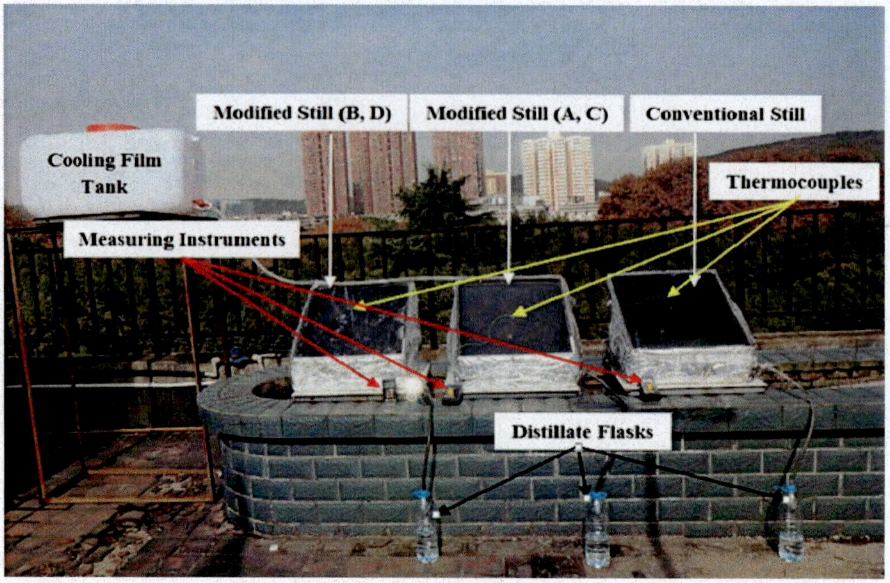

Fig 3.16 Experimental setup for the water solar distiller [103]

70% enhanced in liquid state. The carbon-based nanoparticles also enhanced the TC and thermal performance of solar equipments was also improved. Sharshir et al. [103] modified following alterations as illustrated in Fig. 3.16 in water solar distiller as:

(1) Adding graphite NPs
(2) Combination of graphite NPs and PCMs
(3) Adding graphite NPs along film cooling
(4) Adding graphite NPs, PCM along film cooling.

The results indicated that due to these modifications, the increment in water productivity was 50.28%, 65%, 56.15%, 73.8%, respectively. Thus, the usage of PCMs, nanoparticles and various TC enhancement materials improve the productivity of water.

3.2.2.3 Cascade PCM

The thermal energy can be stored in layers of energy storage materials as PCM layers of various kinds can be generated to form consecutive multi-layer PCM. This multi-layered PCM is called cascade latent heat storage, which reduces the cost and improves efficiency of storage. Figure 3.17 shows cascade shell and tube PCM storage system in which different PCMs are connected in series with melting temperature decreases during charging and increase during discharging from left to right.

The cascade PCM has the following advantages as [112, 113]:

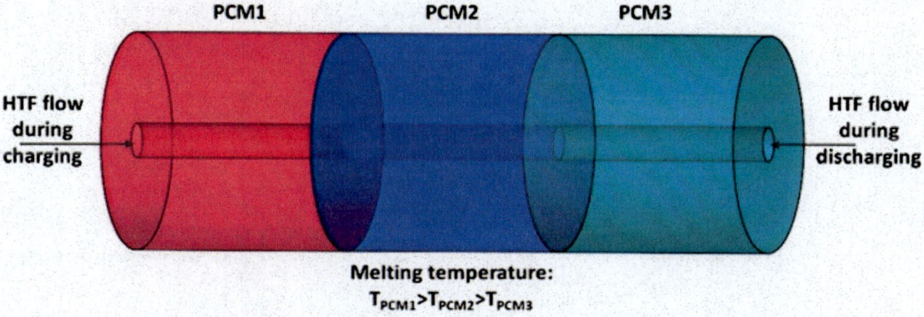

Fig 3.17 The pattern of cascade PCM storage [111]

- The efficiencies of storing capacity enhance by cascade latent heat storage (CLHS)
- The uniform output temperature achieved by CLHS
- The charging and discharging time decrease
- Suitable for high-temperature applications.

Elsanusi and Nsofor [114] investigated the performance of cascade PCM of various configurations for energy storage system by numerical simulations in order to measure conduction and natural convection effects on heat transfer (HT). The authors found that performance was enhanced, and HT effect was interested. The parallel configuration enhanced conduction but convection deceased. The series and vertical directions of cascade PCM reduced melting by 47% and 60%, respectively, compared to single PCM.

Michel and Pitz-Paal [112] studied cascade heat storage having alkali nitrates salts ($NaNO_3$, KNO_3) and performed experiments on shell and tube heat exchanger. The results were used in numerical model under various cascade PCM configurations. The results indicated that cascade latent heat design was suitable for temperature up to 100 °C and was complex for high temperature. However, low TC of these PCMs was a challenging task, which must be overcome in future. Shabgard et al. [113] proposed that when low melting temperature PCM was used then the highest exergy efficiency was achieved. The exergy recovered for 24 h during charging-discharging of cascade PCM was 10% better compared to single PCM.

Domanski and Fallah [115] investigated the exergy behavior of cascade storage and investigated one-stage, three-stage, five-stage storage. The results found that energy and exergy efficiency was greatly improved by using cascade type of storages. Flueckiger and Garimella [116] investigated thermocline cascade storage using finite volume approach, which was connected to solar power plant as shown in Fig. 3.18. It was cleared that by using cascade PCM, storage capacity was enhanced by supplying more amount of steam, but the ability to use thermocline heat storage was decreased.

Ankit Yadav [117] performed experiments by using multiple PCM to evaluate TES capability. The two PCMs, namely, d-mannitol and hydroquinone, with melting temperature range 150–200 °C was selected due to high heat of fusion and efficiency was compared. The results indicated that the effectiveness of setup was improved to

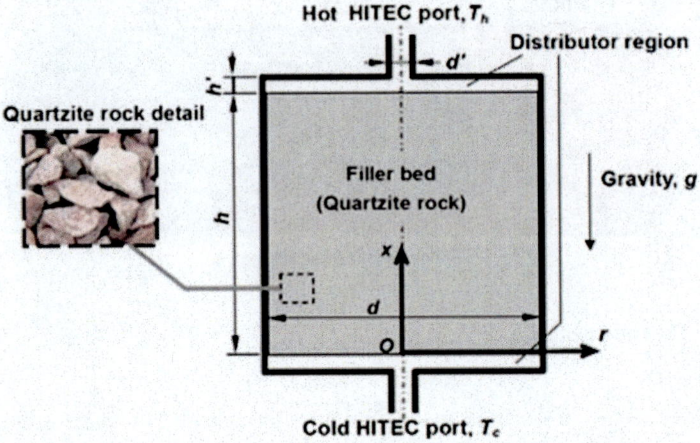

Fig 3.18 Schematic illustration of thermocline tank thermal energy storage [116]

19.36% by using multiple PCMs when compared to single PCM. Shamsi et al. [118] proposed one-dimensional model of packed bed storage, which was verified by using literature experimental data. The results were optimized by comparing with single PCM and results showed that optimal design showed 5.12% increased in storage capacity compared to single PCM. Boyon et al. [119] proposed solid–solid phase change cascade system (shown in Fig. 3.19) having mixture of lithium and sodium sulfate, which can be coupled to solar power plant. Such cascade system has low thermal expansion, low corrosiveness, no need for encapsulation and capable of high temperature to 576 °C. The mixture of 59.17% of $NaLiSO_4$ and 40.83% of Li_2SO_4 gave optimum charge of both PCM for power cycles. The economic analysis showed that cascaded system had cost about \$50.4 $KWh_{th}{}^{-1}$.

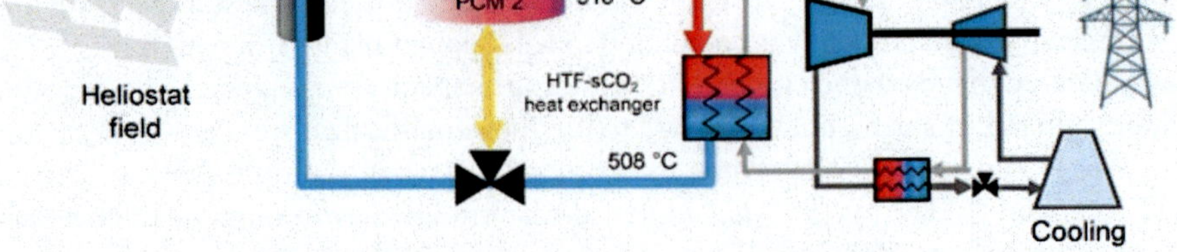

Fig 3.19 Proposed solar thermal power plant with cascaded PCM storage for power cycle [119]

Tehrani et al. [120] investigated the performance of thermal energy storage integrated with 19.9 MW Gemasolar concentrated solar power plant located in Spain and compared with traditional two-tank system. The performance of single, cascade and multilayered thermal energy storage system was evaluated. The results showed that there was a significantly difference between charging and discharging time of cascade and single PCM and high melting temperature of PCM showed better performance. Yuan et al. [121] studied high-temperature cascade type molten salt to measure charging and discharging effects for latent thermal energy storage. The PCM with various concentrations of 34.83 wt% Li_2CO_3-65.17 wt% K_2CO_3, 45.59 wt% Li_2CO_3-53.41 wt% K_2CO_3, 22 wt% Li_2CO_3-62.0 wt% K_2CO_3, 16.0 wt% Na_2CO_3 were used in the experiments. The results indicated that charging and discharging time of cascade PCM was enhanced by 38.40% and 27.02% compared to non-cascaded PCM. Peiro et al. [122] compared the performance of single and cascaded PCM with melting temperature range 150–200 °C. The cascaded PCM showed the effectiveness increment of 19.36% and cascaded configuration exhibited great uniformity in terms of temperature difference. Mao and Zhang [123] studied cascaded PCM in concentrated solar plant to determine the performance for thermal storage capacity. A model based on concentric dispersion model was developed to study phase transition of PCM. The results concluded that storage capacity and utilization of cascaded PCM were higher and it was increased to about 2–3% but, however, this increment was too small. Kim et al. [124] investigated cascade type paraffin wax with different melting points for thermal management of heat sinks along with numerical calculations. It was found that peak temperature of heat sink was decreased from 123.4 to 107.2 °C decreased by 13% due to cascade PCM. Zayed et al. [125] reviewed the cascaded PCMs efficiency for solar water collector tanks for better energy storage. The authors also made comparison of thermal performance improvement, enhancement in HT and thermodynamic optimization of various solar water collected equipped with cascaded PCM. It was concluded that multi-tank integrated with cascaded PCM was best suitable for solar water collector. Beemkumar et al. [126] studied multiple PCMs by using various configuration of encapsulations. The D-mannitol, D-sorbitol and paraffin were used as PCMs and spherical balls of copper, aluminum and brass were used for encapsulation. The comparison was done, and the best combination was found with copper balls and aluminum balls were best in term of cost.

3.2.3 Hybrid PCM

The different researchers have also done work to enhance the value of "k" by making hybrid PCM contained organic–inorganic materials. These hybrid PCMs can be prepared by doping NPs in these materials. These NPs have a large value of "k" results in enhancement in performance of hybrid PCM [127–129]. These hybrid PCMs are then used in various applications like solar energy, energy storage and in buildings. The various hybrid PCMs developed by different researchers are represented in Fig. 3.20.

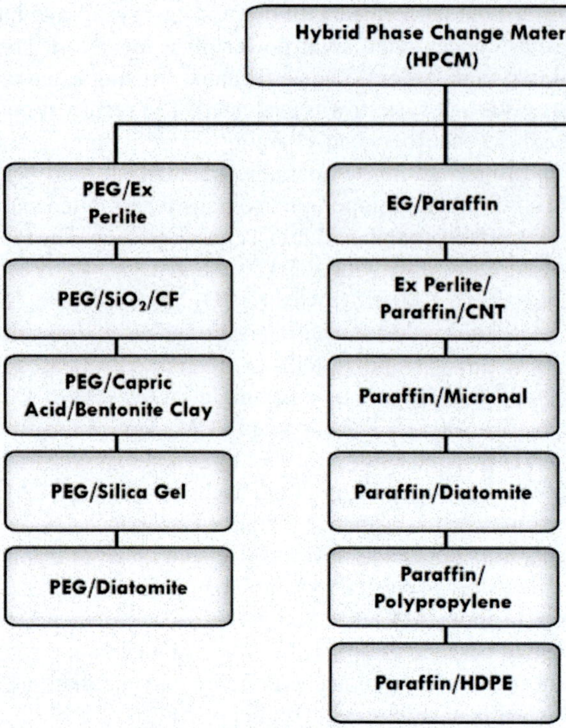

Fig 3.20 Different hybrid PCMs prepared by various researchers [130]

Zhang et al. [131] prepared hybrid PCM by introducing tetrabromobisphenol-A and decabrominated-dipheny ethane at various concentrations in polyethylene glycol/toluene PCM. The thermal characteristics were determined, and results showed that resulted hybrid PCM indicated high phase change enthalpy and also this hybrid PCM was good fire resistant as well as high thermal energy storage capacity. Su et al. [132] also prepared hybrid PCM by adding 10 wt% of hexagonal boron nitride in n-octadecane and steric acid. The results depicted that the value of "k" of prepared hybrid PCM enhanced by 8% in solid state and 16.7% in melting state by adding hexagonal boron nitride.

Zhang et al. [133] made hybrid PCM using vacuum impregnation method by mixing expended perlite and polyethylene glycol. The process is represented in Fig. 3.21 in which 33.63 wt% of carbon fiber was added to improve the performance of hybrid PCM in thermal storage applications. It was concluded that the value of "k" was enhanced by 2.97 times compared to polyethylene glycol. The prepared composite was also efficient in terms of thermal stability and can be used in various thermal storage applications.

Hong et al. [134] prepared form stable PCM by combining superwetting polypropylene aerogel with paraffin having 1060 wt%. The results indicated that the value of "k" was enhanced two times compared to pure paraffin. Karaipekli

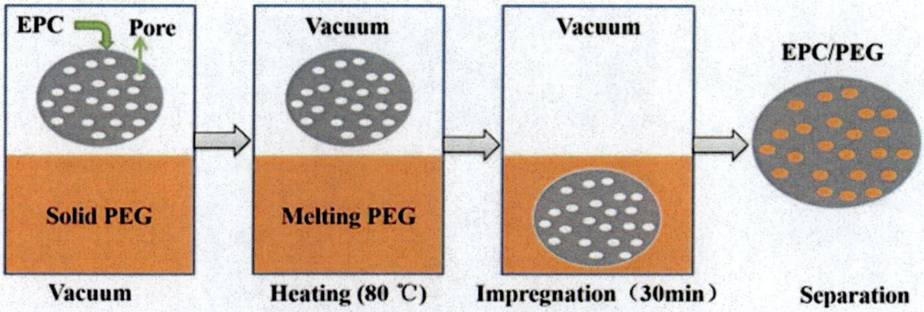

Fig 3.21 Schematic for the preparation of hybrid PCM [133]

et al. [135] measured the thermal properties of expanded perlite/paraffin composite by adding carbon nanotubes having mass fraction of 0.3%, 0.5% and 1%. It was revealed that the value of "k" was enhanced by 113.3% at mass fraction of 1%. Sari [136] prepared shape stabilized PCM of bentonite-based by impregnating 40 wt% of capric acid, 43 wt% of polyethylene glycol, 18 wt% of heptadecane and 32 wt% of dodecanol in bentonite clay. The SEM images of prepared hybrid PCM are shown in Fig. 3.22 and results indicated that latent heat capacity of prepared hybrid PCM was improved, which was later used in building applications.

3.2.4 *MXene-Based PCM*

The name of MXene (pronounced "maxines"), "MX" means that it is derivative from MAX phase materials and "ene" means that it has graphene-like microscopic structure [143]. MXenes are produced as a result of etching of definite atomic layers of MAX layers, which are large group of ternary carbides and nitrides along double transition metal structure [144, 145]. MXenes are mostly represented in general formula $M_{n+1}X_nT_x$ and are derivative from precursor MAX phase $M_{n+1}AX_n$, where M is transition metal, A belongs to elements from groups 12 to 16 of periodic table, X stands to carbon or nitrogen, T donates surface termination group as fluorine, oxygen, chlorine, x indicates number of surface functionalities and n = 1,2 etc. [146]. In MAX phase, A layer is sandwiched within octahedral $M_{n+1}X_n$ having stronger M–X bond and weak M–A bond [147, 148]. The forming elements of MAX phase from periodic table are represented in Fig. 3.23 [149].

MXenes have three different formulas as M_2X, M_3X, M_4X and it can be mono M-element, solution form from two various M-element and ordered doubled M-element. MXenes have crystal structures as well as ordered phase MXenes are higher stable than solid solution [150]. Researchers at Drexel University first fabricated multilayered MXene in 2011, which was fabricated by etching layer of MAX phase of Ti_3AlC_2 [151]. There are more than 70 types of MAX phases [152, 153] present and more way to go. Among these, titanium-based MXenes such as $Ti_3C_2T_x$ and

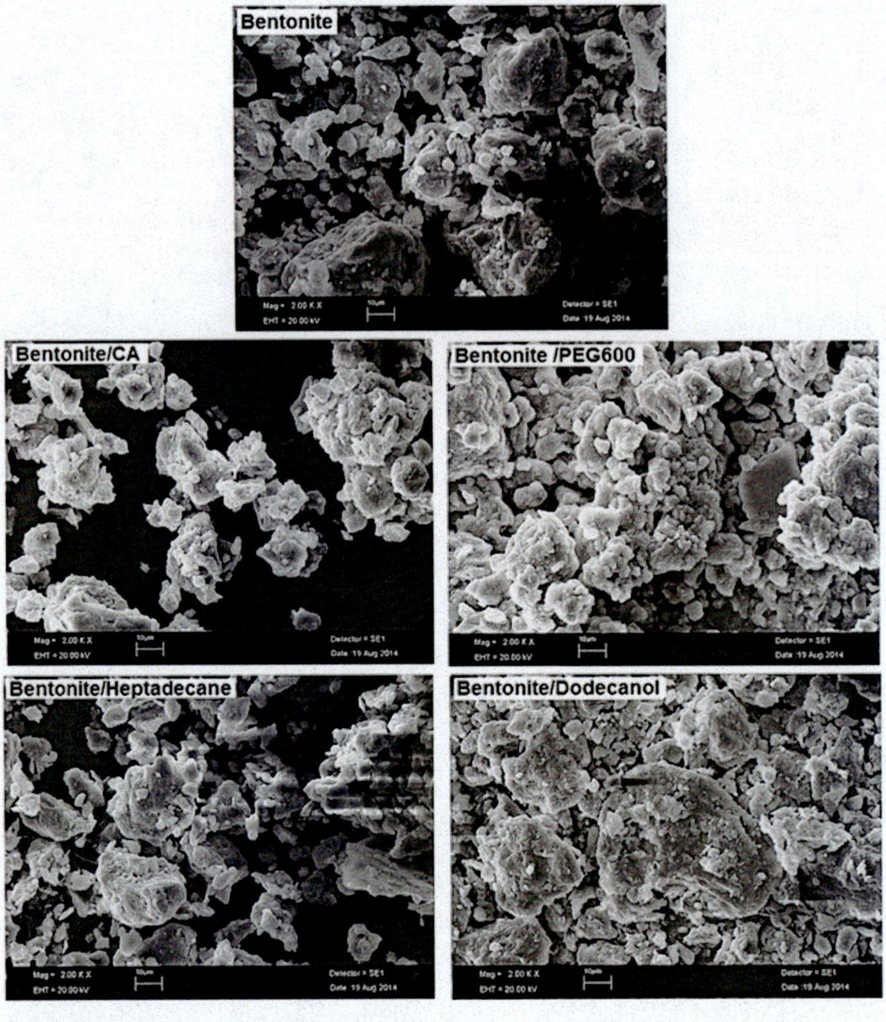

Fig 3.22 SEM images of prepared hybrid PCM [136]

Ti_3CT_x have widely studied by various investigators [154]. Along carbides, other MXene structures having nitrides and carbonitrides were studied also in different applications [155–157]. The different method of preparation for MXene is shown in Fig. 3.24 that includes bottom-up synthesis and top-down technique. The bottom-up synthesis method includes chemical vapor deposition method, template method and plasma-enhanced pulsed method. The top-down synthesis technique contains different procedures like precursors, etching and exfoliation in MXene preparation.

Yang et al. [158] synthesized 3D porous bacterial cellulose with MXene, which were mixed along polyethylene glycol as PCM resulted in the formation of composite PCM having excellent photothermal conversion efficiency as shown in Fig. 3.25. The

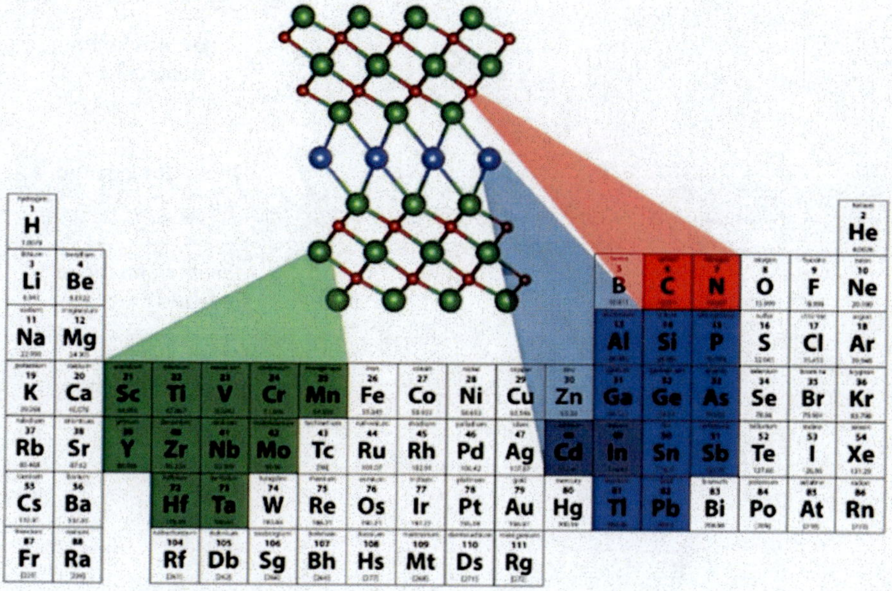

Fig 3.23 Periodic table elements that form MAX phase [149]

composite PCM indicated large storage capacity, good shape stability and efficient photothermal conversion. The composite PCM retained its shape at heating of up to 120 °C without any leakage. The thermal stability and chemical reliability also not changed even after 100 cycles. The storage capacity and photo absorption capacity of composite PCM were excellent making it best for solar energy applications.

This chapter concludes various advanced energy storage materials including molten salts nanofluids, composite PCMs, i.e. nano-based PCM, cascade PCM, hybrid PCM and MXene-based PCM. These advanced energy storage materials have much thermal properties compared to conventional energy storage materials. The value of TC and storage capacity has increased significantly by these materials and has much better performance for energy storage. However, there still some limitations regarding thermal stability, storage capacity and formation technique. Also, appropriate type of selection of PCM is necessary according to the application. These kinds of issues must be addressed in future research and suitable selection of PCM is compulsory for making it better in performance in terms of storage capacity and stability.

Molten salt-based nanofluids have great potential for HT and storage of molten salts but considerable advancement in this field is still necessary. The stability and size of NPs in molten salts, its characterization and corrosion issues should be considered for future research. Encapsulation of PCM results in better performance of base PCM. The encapsulation of PCM is the promising solution to minimize the leakage

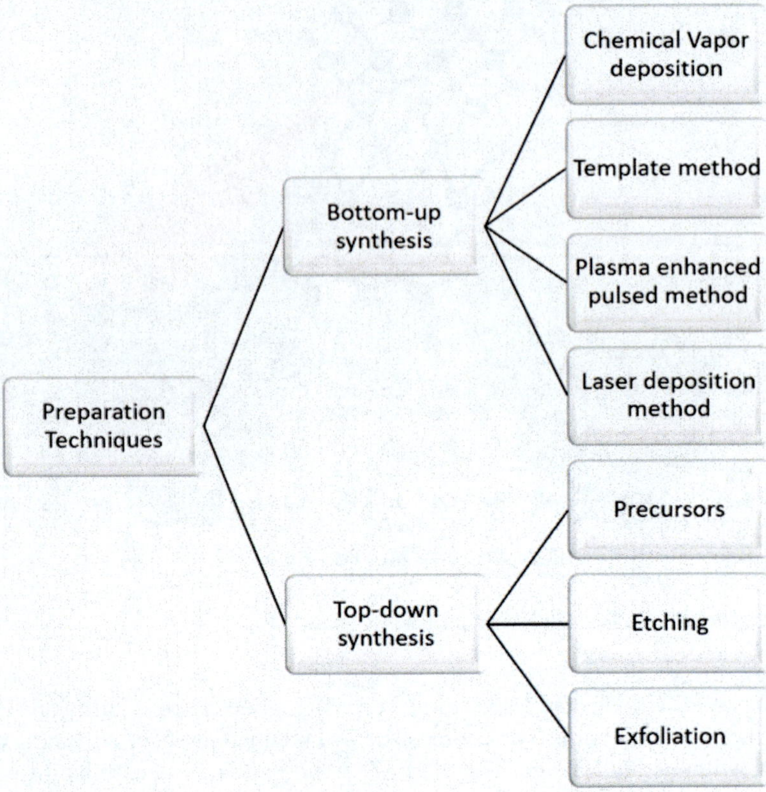

Fig 3.24 Various methods for the preparation of MXene [158]

of PCM along subcooling and supercooling issues but standard testing measurement must be undertaken for leakage testing. The standard stability testing must also need to be considered after final formation of capsulated PCM. The researchers should focus on increment in strength of capsule materials. The addition of metallic or non-metallic NPs energized storage capacity of energy storage materials but they face huge difficulties for energy storage on energy system. But the summarized study concluded that heat transfer capability and thermal energy storage capacity increased. Cascade PCMs have promising applications for thermal energy storage and increase stage results in enhancement of HT and storage capacity compared to single stage PCM. However, very limited experimental studies are found regarding cascade thermal storage. Hybrid- and MXene-based PCM are a new class of advance energy storage materials and a lot of research is still needed to improve the performance of these materials.

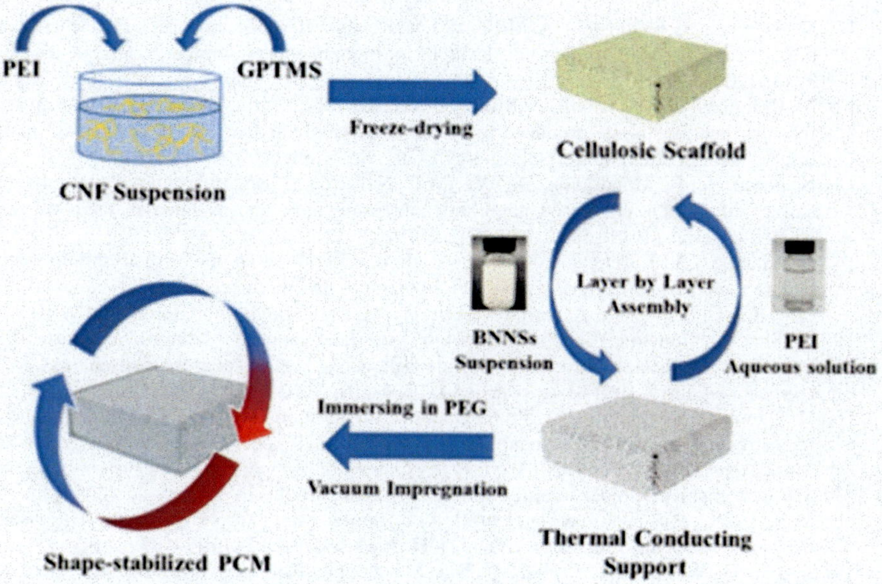

Fig 3.25 Schematic representation of the preparation of PEG-based shape stabilized PCM [158]

References

1. V.M.B. Nunes, C.S. Queirós, M.J.V. Lourenço, F.J. V Santos, C.A.N. De Castro, Molten salts as engineering fluids–a review: part I. Molten alkali nitrates. Appl. Energy. **183**, 603–611 (2016)
2. K. Furukava, *Fundamentals of Molten Salt Thermal Technology* (1980)
3. V.M.B. Nunes, F.J.V. Santos, M. Lopes, C.A.N. de Castro, Accurate measurement of physicochemical properties on ionic liquids and molten salts, molten salts ion. Liq. Never Twain 229 (2010)
4. T.R. Shah, H.M. Ali, Applications of hybrid nanofluids in solar energy, practical limitations and challenges: a critical review. Sol. Energy **183**, 173–203 (2019)
5. R. Jacob, M. Belusko, A.I. Fernández, L.F. Cabeza, W. Saman, F. Bruno, Embodied energy and cost of high temperature thermal energy storage systems for use with concentrated solar power plants. Appl. Energy **180**, 586–597 (2016)
6. A. Gil, M. Medrano, I. Martorell, A. Lázaro, P. Dolado, B. Zalba, L.F. Cabeza, State of the art on high temperature thermal energy storage for power generation. Part 1—concepts, materials and modellization. Renew. Sustain. Energy Rev. **14**, 31–55 (2010)
7. S.M. Flueckiger, Z. Yang, S.V. Garimella, Review of molten-salt thermocline tank modeling for solar thermal energy storage. Heat Transf. Eng. **34**, 787–800 (2013)
8. U. Herrmann, B. Kelly, H. Price, Two-tank molten salt storage for parabolic trough solar power plants. Energy **29**, 883–893 (2004)
9. R. Devaradjane, Utilization of molten nitrate salt nanomaterials for heat capacity enhancement in solar power applications (2013)
10. B. Muñoz-Sánchez, J. Nieto-Maestre, I. Iparraguirre-Torres, A. García-Romero, J.M. Sala-Lizarraga, Molten salt-based nanofluids as efficient heat transfer and storage materials at high temperatures. An overview of the literature, Renew. Sustain. Energy Rev. **82**, 3924–3945 (2018)

11. M. Chieruzzi, G.F. Cerritelli, A. Miliozzi, J.M. Kenny, Effect of nanoparticles on heat capacity of nanofluids based on molten salts as PCM for thermal energy storage. Nanoscale Res. Lett. **8**, 448 (2013)
12. M. Chieruzzi, G.F. Cerritelli, A. Miliozzi, J.M. Kenny, L. Torre, Heat capacity of nanofluids for solar energy storage produced by dispersing oxide nanoparticles in nitrate salt mixture directly at high temperature. Sol. Energy Mater. Sol. Cells. **167**, 60–69 (2017)
13. N. Navarrete, R. Mondragón, D. Wen, M.E. Navarro, Y. Ding, J.E. Juliá, Thermal energy storage of molten salt–based nanofluid containing nano-encapsulated metal alloy phase change materials. Energy **167**, 912–920 (2019)
14. M. Chieruzzi, J.M. Kenny, A. Miliozzi, Studio e sviluppo di un mezzo di accumulo a calore latente a media temperatura (200–400 C) costituito da una miscela di sali e nanoparticelle. Rep. Ric. Di Sist. Elettr. Accord. Di Program. Minist. Dello Svilupp. Econ. **98** (2013)
15. D. Shin, D. Banerjee, Enhancement of specific heat capacity of high-temperature silica-nanofluids synthesized in alkali chloride salt eutectics for solar thermal-energy storage applications. Int. J. Heat Mass Transf. **54**, 1064–1070 (2011)
16. H. Tiznobaik, D. Banerjee, D. Shin, Effect of formation of "long range" secondary dendritic nanostructures in molten salt nanofluids on the values of specific heat capacity. Int. J. Heat Mass Transf. **91**, 342–346 (2015)
17. V.V.B. Somani, Colloidal stability of magnetic nanoparticles in molten salts (2010)
18. M.X. Ho, C. Pan, Optimal concentration of alumina nanoparticles in molten Hitec salt to maximize its specific heat capacity. Int. J. Heat Mass Transf. **70**, 174–184 (2014)
19. S. Liu, D. Wu, J. Liu, Y. Nian, P. Qiu, Development of a novel molten-salt filled with nanoparticles for concentration solar plants (2013)
20. M. Lasfargues, Nitrate based high temperature nano-heat-transfer-fluids: formulation & characterisation (2014)
21. A.G. Fernández, B. Muñoz-Sánchez, J. Nieto-Maestre, A. García-Romero, High temperature corrosion behavior on molten nitrate salt-based nanofluids for CSP plants. Renew. Energy. **130**, 902–909 (2019)
22. U. Nithiyanantham, A. Zaki, Y. Grosu, L. González-Fernández, J.M. Igartua, A. Faik, SiO2@ Al2O3 core-shell nanoparticles based molten salts nanofluids for thermal energy storage applications. J. Energy Storage **26**, 101033 (2019)
23. Y. Li, X. Chen, Y. Wu, Y. Lu, R. Zhi, X. Wang, C. Ma, Experimental study on the effect of SiO2 nanoparticle dispersion on the thermophysical properties of binary nitrate molten salt. Sol. Energy **183**, 776–781 (2019)
24. A. Svobodova-Sedlackova, C. Barreneche, G. Alonso, A.I. Fernandez, P. Gamallo, Effect of nanoparticles in molten salts–MD simulations and experimental study. Renew. Energy **152**, 208–216 (2020)
25. Y. Grosu, U. Nithiyanantham, L. González-Fernández, A. Faik, Preparation and characterization of nanofluids based on molten salts with enhanced thermophysical properties for thermal energy storage at concentrate solar power, in: *AIP Conference on Proceedings*. AIP Publishing LLC (2019), p. 200021
26. J. Nieto-Maestre, B. Muñoz-Sánchez, A.G. Fernández, A. Faik, Y. Grosu, A. García-Romero, Compatibility of container materials for Concentrated Solar Power with a solar salt and alumina based nanofluid: A study under dynamic conditions. Renew. Energy **146**, 384–396 (2020)
27. X. Wei, Y. Yin, B. Qin, W. Wang, J. Ding, J. Lu, Preparation and enhanced thermal conductivity of molten salt nanofluids with nearly unaltered viscosity. Renew. Energy **145**, 2435–2444 (2020)
28. M.N.A. Hawlader, M.S. Uddin, M.M. Khin, Microencapsulated PCM thermal-energy storage system. Appl. Energy **74**, 195–202 (2003)
29. W. Aftab, X. Huang, W. Wu, Z. Liang, A. Mahmood, R. Zou, Nanoconfined phase change materials for thermal energy applications. Energy Environ. Sci. **11**, 1392–1424 (2018)
30. B. Chen, X. Wang, R. Zeng, Y. Zhang, X. Wang, J. Niu, Y. Li, H. Di, An experimental study of convective heat transfer with microencapsulated phase change material suspension: laminar flow in a circular tube under constant heat flux. Exp. Therm. Fluid Sci. **32**, 1638–1646 (2008)

31. C. Alkan, A. Sarı, A. Karaipekli, O. Uzun, Preparation, characterization, and thermal properties of microencapsulated phase change material for thermal energy storage. Sol. Energy Mater. Sol. Cells. **93**, 143–147 (2009)
32. Y. Konuklu, M. Ostry, H.O. Paksoy, P. Charvat, Review on using microencapsulated phase change materials (PCM) in building applications. Energy Build. **106**, 134–155 (2015)
33. A. Jamekhorshid, S.M. Sadrameli, M. Farid, A review of microencapsulation methods of phase change materials (PCMs) as a thermal energy storage (TES) medium. Renew. Sustain. Energy Rev. **31**, 531–542 (2014)
34. H. Zhang, X. Wang, D. Wu, Silica encapsulation of n-octadecane via sol–gel process: a novel microencapsulated phase-change material with enhanced thermal conductivity and performance. J. Colloid Interface Sci. **343**, 246–255 (2010)
35. A. Arshad, M. Jabbal, Y. Yan, J. Darkwa, The micro-/nano-PCMs for thermal energy storage systems: a state of art review. Int. J. Energy Res. **43**, 5572–5620 (2019)
36. Y.E. Milian, A. Gutierrez, M. Grageda, S. Ushak, A review on encapsulation techniques for inorganic phase change materials and the influence on their thermophysical properties. Renew. Sustain. Energy Rev. **73**, 983–999 (2017)
37. Y. Zhang, Z. Chen, Q. Wang, Q. Wu, Melting in an enclosure with discrete heating at a constant rate. Exp. Therm. Fluid Sci. **6**, 196–201 (1993)
38. A.F. Regin, S.C. Solanki, J.S. Saini, Latent heat thermal energy storage using cylindrical capsule: numerical and experimental investigations. Renew. Energy **31**, 2025–2041 (2006)
39. K. Yuan, H. Wang, J. Liu, X. Fang, Z. Zhang, Novel slurry containing graphene oxide-grafted microencapsulated phase change material with enhanced thermo-physical properties and photo-thermal performance. Sol. Energy Mater. Sol. Cells. **143**, 29–37 (2015)
40. S. Yu, X. Wang, D. Wu, Microencapsulation of n-octadecane phase change material with calcium carbonate shell for enhancement of thermal conductivity and serving durability: synthesis, microstructure, and performance evaluation. Appl. Energy **114**, 632–643 (2014)
41. L. Chai, X. Wang, D. Wu, Development of bifunctional microencapsulated phase change materials with crystalline titanium dioxide shell for latent-heat storage and photocatalytic effectiveness. Appl. Energy **138**, 661–674 (2015)
42. M. Safdari, R. Ahmadi, S. Sadeghzadeh, Numerical investigation on PCM encapsulation shape used in the passive-active battery thermal management. Energy **193**, 116840 (2020)
43. B. Praveen, S. Suresh, Thermal performance of micro-encapsulated PCM with LMA thermal percolation in TES based heat sink application. Energy Convers. Manag. **185**, 75–86 (2019)
44. S. Seitz, H. Ajiro, Self-assembling weak polyelectrolytes for the layer-by-layer encapsulation of paraffin-type phase change material icosane. Sol. Energy Mater. Sol. Cells. **190**, 57–64 (2019)
45. G.-Q. Zhanga, C.-W. Caia, G.-C. Zhua, L. Zhoua, G.-J. Liua, J. Militkyb, J. Marekc, M. Venkataramanb, J.-M. Yaoa, Preparation, properties and encapsulation of high thermostability phase-change material. J. Fiber Bioeng. Inf. **12**, 43–54 (2019)
46. H. Nikpourian, A.R. Bahramian, M. Abdollahi, On the thermal performance of a novel PCM nanocapsule: the effect of core/shell. Renew. Energy **151**, 322–331 (2020)
47. C.-J. Ho, W.-L. Chou, C.-M. Lai, Thermal and electrical performance of a water-surface floating PV integrated with a water-saturated MEPCM layer. Energy Convers. Manag. **89**, 862–872 (2015)
48. L. Qiu, Y. Ouyang, Y. Feng, X. Zhang, Review on micro/nano phase change materials for solar thermal applications. Renew. Energy **140**, 513–538 (2019)
49. J.M. Khodadadi, S.F. Hosseinizadeh, Nanoparticle-enhanced phase change materials (NEPCM) with great potential for improved thermal energy storage. Int. Commun. Heat Mass Transf. **34**, 534–543 (2007)
50. K.Y. Leong, M.R.A. Rahman, B.A. Gurunathan, Nano-enhanced phase change materials: a review of thermo-physical properties, applications and challenges. J. Energy Storage **21**, 18–31 (2019)

51. J.M. Khodadadi, L. Fan, Expedited freezing of nanoparticle-enhanced phase change materials (NEPCM) exhibited through a simple 1-D Stefan problem formulation, in: *ASME 2009 Heat Transfer Summer Conference Collocated with InterPACK09 3rd Energy Sustainability Conferences*. American Society of Mechanical Engineers (2009), pp. 345–351
52. T. Qian, J. Li, X. Min, W. Guan, Y. Deng, L. Ning, Enhanced thermal conductivity of PEG/diatomite shape-stabilized phase change materials with Ag nanoparticles for thermal energy storage. J. Mater. Chem. A **3**, 8526–8536 (2015)
53. S. Wu, D. Zhu, X. Zhang, J. Huang, Preparation and melting/freezing characteristics of Cu/paraffin nanofluid as phase-change material (PCM). Energy Fuels **24**, 1894–1898 (2010)
54. J.L. Zeng, Z. Cao, D.W. Yang, L.X. Sun, L. Zhang, Thermal conductivity enhancement of Ag nanowires on an organic phase change material. J. Therm. Anal. Calorim. **101**, 385–389 (2010)
55. H.H. Al-Kayiem, S.C. Lin, Performance evaluation of a solar water heater integrated with a PCM nanocomposite TES at various inclinations. Sol. Energy **109**, 82–92 (2014)
56. A.A.A. Attia, A.A. Altohamy, M.F.A. Rabbo, R.Y. Sakr, Comparative study on AL2O3 nanoparticle addition on cool storage system performance. Appl. Therm. Eng. **94**, 449–457 (2016)
57. A.A. Altohamy, M.F.A. Rabbo, R.Y. Sakr, A.A.A. Attia, Effect of water based Al2O3 nanoparticle PCM on cool storage performance. Appl. Therm. Eng. **84**, 331–338 (2015)
58. A.O. Elsayed, Numerical study on performance enhancement of solid–solid phase change materials by using multi-nanoparticles mixtures. J. Energy Storage **4**, 106–112 (2015)
59. A.A.R. Darzi, M. Jourabian, M. Farhadi, Melting and solidification of PCM enhanced by radial conductive fins and nanoparticles in cylindrical annulus. Energy Convers. Manag. **118**, 253–263 (2016)
60. J.M. Mahdi, E.C. Nsofor, Melting of PCM with nanoparticles in a triplex-tube thermal energy storage system. ASHRAE Trans. **122**, 215–225 (2016)
61. A. Babapoor, G. Karimi, M. Khorram, Fabrication and characterization of nanofiber-nanoparticle-composites with phase change materials by electrospinning. Appl. Therm. Eng. **99**, 1225–1235 (2016)
62. L. Colla, L. Fedele, S. Mancin, L. Danza, O. Manca, Nano-PCMs for enhanced energy storage and passive cooling applications. Appl. Therm. Eng. **110**, 584–589 (2017)
63. S. Motahar, N. Nikkam, A.A. Alemrajabi, R. Khodabandeh, M.S. Toprak, M. Muhammed, A novel phase change material containing mesoporous silica nanoparticles for thermal storage: a study on thermal conductivity and viscosity. Int. Commun. Heat Mass Transf. **56**, 114–120 (2014)
64. R.K. Sharma, P. Ganesan, V.V. Tyagi, H.S.C. Metselaar, S.C. Sandaran, Thermal properties and heat storage analysis of palmitic acid-TiO2 composite as nano-enhanced organic phase change material (NEOPCM). Appl. Therm. Eng. **99**, 1254–1262 (2016)
65. M.S.M. Al-Jethelah, S.H. Tasnim, S. Mahmud, A. Dutta, Melting of nano-phase change material inside a porous enclosure. Int. J. Heat Mass Transf. **102**, 773–787 (2016)
66. S.K. Sahoo, M.K. Das, P. Rath, Numerical study of cyclic melting and solidification of nano enhanced phase change material based heat sink in thermal management of electronic components, in: *ASME 2016 5th International Conference on Micro/Nanoscale Heat Mass Transformation*. American Society of Mechanical Engineers (2016), pp. V002T10A005–V002T10A005
67. R. Hossain, S. Mahmud, A. Dutta, I. Pop, Energy storage system based on nanoparticle-enhanced phase change material inside porous medium. Int. J. Therm. Sci. **91**, 49–58 (2015)
68. M. Abdollahzadeh, M. Esmaeilpour, Enhancement of phase change material (PCM) based latent heat storage system with nano fluid and wavy surface. Int. J. Heat Mass Transf. **80**, 376–385 (2015)
69. M. Jourabian, M. Farhadi, K. Sedighi, On the expedited melting of phase change material (PCM) through dispersion of nanoparticles in the thermal storage unit. Comput. Math. Appl. **67**, 1358–1372 (2014)

70. M. Scarselli, P. Castrucci, M. De Crescenzi, Electronic and optoelectronic nano-devices based on carbon nanotubes. J. Phys. Condens. Matter. **24**, 313202 (2012)
71. H. Dai, Carbon nanotubes: opportunities and challenges. Surf. Sci. **500**, 218–241 (2002)
72. R. Sengupta, M. Bhattacharya, S. Bandyopadhyay, A.K. Bhowmick, A review on the mechanical and electrical properties of graphite and modified graphite reinforced polymer composites. Prog. Polym. Sci. **36**, 638–670 (2011)
73. C. Amaral, R. Vicente, P. Marques, A. Barros-Timmons, Phase change materials and carbon nanostructures for thermal energy storage: a literature review. Renew. Sustain. Energy Rev. **79**, 1212–1228 (2017)
74. W. Yu, S.U.S. Choi, The role of interfacial layers in the enhanced thermal conductivity of nanofluids: a renovated Hamilton-Crosser model. J. Nanoparticle Res. **6**, 355–361 (2004)
75. H. Taherian, J.L. Alvarado, E.M. Languri, Enhanced thermophysical properties of multiwalled carbon nanotubes based nanofluids. Part 1: critical review. Renew. Sustain. Energy Rev. **82**, 4326–4336 (2018)
76. M.J. Assael, I.N. Metaxa, J. Arvanitidis, D. Christofilos, C. Lioutas, Thermal conductivity enhancement in aqueous suspensions of carbon multi-walled and double-walled nanotubes in the presence of two different dispersants. Int. J. Thermophys. **26**, 647–664 (2005)
77. K.W. Shah, A review on enhancement of phase change materials–a nanomaterials perspective. Energy Build. (2018)
78. Y. Cui, C. Liu, S. Hu, X. Yu, The experimental exploration of carbon nanofiber and carbon nanotube additives on thermal behavior of phase change materials. Sol. Energy Mater. Sol. Cells. **95**, 1208–1212 (2011)
79. J. Wang, H. Xie, Z. Xin, Y. Li, Increasing the thermal conductivity of palmitic acid by the addition of carbon nanotubes. Carbon N. Y. **48**, 3979–3986 (2010)
80. J. Wang, H. Xie, Z. Xin, Y. Li, L. Chen, Enhancing thermal conductivity of palmitic acid based phase change materials with carbon nanotubes as fillers. Sol. Energy **84**, 339–344 (2010)
81. Q. Tang, J. Sun, S. Yu, G. Wang, Improving thermal conductivity and decreasing supercooling of paraffin phase change materials by n-octadecylamine-functionalized multi-walled carbon nanotubes. RSC Adv. **4**, 36584–36590 (2014)
82. P.M. Gilart, Á.Y. Martínez, M.G. Barriuso, C.M. Martínez, Development of PCM/carbon-based composite materials. Sol. Energy Mater. Sol. Cells. **107**, 205–211 (2012)
83. J.J. Vilatela, R. Khare, A.H. Windle, The hierarchical structure and properties of multifunctional carbon nanotube fibre composites. Carbon N. Y. **50**, 1227–1234 (2012)
84. M. Xing, J. Yu, R. Wang, Experimental study on the thermal conductivity enhancement of water based nanofluids using different types of carbon nanotubes. Int. J. Heat Mass Transf. **88**, 609–616 (2015)
85. L. Chen, R. Zou, W. Xia, Z. Liu, Y. Shang, J. Zhu, Y. Wang, J. Lin, D. Xia, A. Cao, Electro- and photodriven phase change composites based on wax-infiltrated carbon nanotube sponges. ACS Nano **6**, 10884–10892 (2012)
86. Mj. Assael, C.-F. Chen, I. Metaxa, W.A. Wakeham, Thermal conductivity of suspensions of carbon nanotubes in water. Int. J. Thermophys. **25**, 971–985 (2004)
87. M.-S. Liu, M.C.-C. Lin, I.-T. Huang, C.-C. Wang, Enhancement of thermal conductivity with carbon nanotube for nanofluids. Int. Commun. Heat Mass Transf. **32**, 1202–1210 (2005)
88. Y. Ding, H. Alias, D. Wen, R.A. Williams, Heat transfer of aqueous suspensions of carbon nanotubes (CNT nanofluids). Int. J. Heat Mass Transf. **49**, 240–250 (2006)
89. X. Wang, Y. He, G. Cheng, L. Shi, X. Liu, J. Zhu, Direct vapor generation through localized solar heating via carbon-nanotube nanofluid. Energy Convers. Manag. **130**, 176–183 (2016)
90. M. Hashemi, S.H. Noie, Study of flow boiling heat transfer characteristics of critical heat flux using carbon nanotubes and water nanofluid. J. Therm. Anal. Calorim. **130**, 2199–2209 (2017)
91. I.M. Mahbubul, M.M.A. Khan, N.I. Ibrahim, H.M. Ali, F.A. Al-Sulaiman, R. Saidur, Carbon nanotube nanofluid in enhancing the efficiency of evacuated tube solar collector. Renew. Energy **121**, 36–44 (2018)

92. D. Zou, X. Ma, X. Liu, P. Zheng, Y. Hu, Thermal performance enhancement of composite phase change materials (PCM) using graphene and carbon nanotubes as additives for the potential application in lithium-ion power battery. Int. J. Heat Mass Transf. **120**, 33–41 (2018)
93. R. Vajtai, *Springer Handbook of Nanomaterials* (Springer, 2013)
94. T. Nomura, K. Tabuchi, C. Zhu, N. Sheng, S. Wang, T. Akiyama, High thermal conductivity phase change composite with percolating carbon fiber network. Appl. Energy **154**, 678–685 (2015)
95. Z.-T. Yu, X. Fang, L.-W. Fan, X. Wang, Y.-Q. Xiao, Y. Zeng, X. Xu, Y.-C. Hu, K.-F. Cen, Increased thermal conductivity of liquid paraffin-based suspensions in the presence of carbon nano-additives of various sizes and shapes. Carbon N. Y. **53**, 277–285 (2013)
96. E. Ismar, A.S. Sarac, Electrospun polyacrylonitrile–lauric acid composite nanofiber webs as a thermal energy storage material. J. Eng. Fiber. Fabr. **14**, 1558925018824890 (2019)
97. P. Lu, W. Chen, M. Zhu, S. Murray, Embedding lauric acid into polystyrene nanofibers to make high-capacity membranes for efficient thermal energy storage. ACS Sustain. Chem. Eng. **5**, 7249–7259 (2017)
98. Z. Said, H. Zeyad, T.I. Eisa, M.E.L.H. Assad, Nano-enhanced PCM for energy storage, in: 2019 *Advanced Science, Engineering and Technology Conference*. IEEE, pp. 1–6 (2019)
99. K. Kant, A. Shukla, A. Sharma, P.H. Biwole, Heat transfer study of phase change materials with graphene nano particle for thermal energy storage. Sol. Energy **146**, 453–463 (2017)
100. B.R. Sushobhan, S.P. Kar, Thermal modeling of melting of nano based phase change material for improvement of thermal energy storage. Energy Proc. **109**, 385–392 (2017)
101. W. Liang, G. Zhang, H. Sun, P. Chen, Z. Zhu, A. Li, Graphene–nickel/n-carboxylic acids composites as form-stable phase change materials for thermal energy storage. Sol. Energy Mater. Sol. Cells. **132**, 425–430 (2015)
102. R.J. Warzoha, R.M. Weigand, A.S. Fleischer, Temperature-dependent thermal properties of a paraffin phase change material embedded with herringbone style graphite nanofibers. Appl. Energy **137**, 716–725 (2015)
103. S.W. Sharshir, G. Peng, L. Wu, F.A. Essa, A.E. Kabeel, N. Yang, The effects of flake graphite nanoparticles, phase change material, and film cooling on the solar still performance. Appl. Energy **191**, 358–366 (2017)
104. J. Wang, H. Xie, Z. Xin, Y. Li, C. Yin, Investigation on thermal properties of heat storage composites containing carbon fibers. J. Appl. Phys. **110**, 94302 (2011)
105. D.H. Choi, J. Lee, H. Hong, Y.T. Kang, Thermal conductivity and heat transfer performance enhancement of phase change materials (PCM) containing carbon additives for heat storage application. Int. J. Refrig. **42**, 112–120 (2014)
106. W.G. Alshaer, S.A. Nada, M.A. Rady, C. Le Bot, E.P. Del Barrio, Numerical investigations of using carbon foam/PCM/Nano carbon tubes composites in thermal management of electronic equipment. Energy Convers. Manag. **89**, 873–884 (2015)
107. S. Harish, D. Orejon, Y. Takata, M. Kohno, Enhanced thermal conductivity of phase change nanocomposite in solid and liquid state with various carbon nano inclusions. Appl. Therm. Eng. **114**, 1240–1246 (2017)
108. M. Amin, N. Putra, E.A. Kosasih, E. Prawiro, R.A. Luanto, T.M.I. Mahlia, Thermal properties of beeswax/graphene phase change material as energy storage for building applications. Appl. Therm. Eng. **112**, 273–280 (2017)
109. S. Harish, D. Orejon, Y. Takata, M. Kohno, Thermal conductivity enhancement of lauric acid phase change nanocomposite with graphene nanoplatelets. Appl. Therm. Eng. **80**, 205–211 (2015)
110. M. Li, A nano-graphite/paraffin phase change material with high thermal conductivity. Appl. Energy **106**, 25–30 (2013)
111. M. Liu, N.H.S. Tay, S. Bell, M. Belusko, R. Jacob, G. Will, W. Saman, F. Bruno, Review on concentrating solar power plants and new developments in high temperature thermal energy storage technologies. Renew. Sustain. Energy Rev. **53**, 1411–1432 (2016)
112. H. Michels, R. Pitz-Paal, Cascaded latent heat storage for parabolic trough solar power plants. Sol. Energy **81**, 829–837 (2007)

113. H. Shabgard, C.W. Robak, T.L. Bergman, A. Faghri, Heat transfer and exergy analysis of cascaded latent heat storage with gravity-assisted heat pipes for concentrating solar power applications. Sol. Energy **86**, 816–830 (2012)
114. O.S. Elsanusi, E.C. Nsofor, *A Study on the Thermal Energy Storage System Using Multiple PCMs* (2019)
115. R. Domański, G. Fellah, Exergy analysis for the evaluation of a thermal storage system employing PCMs with different melting temperatures. Appl. Therm. Eng. **16**, 907–919 (1996)
116. S.M. Flueckiger, S.V. Garimella, Second-law analysis of molten-salt thermal energy storage in thermoclines. Sol. Energy **86**, 1621–1631 (2012)
117. A. Yadav, *Performance Analysis of Multiple Phase Change Materials Based Thermal Energy Storage System* (2019)
118. H. Shamsi, M. Boroushaki, H. Geraei, Performance evaluation and optimization of encapsulated cascade PCM thermal storage. J. Energy Storage **11**, 64–75 (2017)
119. A. Bayon, M. Liu, D. Sergeev, M. Grigore, F. Bruno, M. Müller, Novel solid–solid phase-change cascade systems for high-temperature thermal energy storage. Sol. Energy **177**, 274–283 (2019)
120. S.S.M. Tehrani, Y. Shoraka, K. Nithyanandam, R.A. Taylor, Cyclic performance of cascaded and multi-layered solid-PCM shell-and-tube thermal energy storage systems: a case study of the 19.9 MWe Gemasolar CSP plant. Appl. Energy **228**, 240–253 (2018)
121. F. Yuan, M.-J. Li, Z. Ma, B. Jin, Z. Liu, Experimental study on thermal performance of high-temperature molten salt cascaded latent heat thermal energy storage system. Int. J. Heat Mass Transf. **118**, 997–1011 (2018)
122. G. Peiró, J. Gasia, L. Miró, L.F. Cabeza, Experimental evaluation at pilot plant scale of multiple PCMs (cascaded) vs. single PCM configuration for thermal energy storage. Renew. Energy **83**, 729–736 (2015)
123. Q. Mao, Y. Zhang, Thermal energy storage performance of a three-PCM cascade tank in a high-temperature packed bed system. Renew. Energy (2020)
124. S.-H. Kim, C.S. Heu, D.R. Kim, S.-W. Kang, Numerical modeling and experimental validation of a phase change material-based compact cascade cooling system for enhanced thermal management. Appl. Therm. Eng. **164**, 114470 (2020)
125. M.E. Zayed, J. Zhao, A.H. Elsheikh, F.A. Hammad, L. Ma, Y. Du, A.E. Kabeel, S.M. Shalaby, Applications of cascaded phase change materials in solar water collector storage tanks: a review. Sol. Energy Mater. Sol. Cells. **199**, 24–49 (2019)
126. B. Nagappan, K. Alagu, Y. Devarajan, Heat transfer enhancement of a cascaded thermal energy storage system with various encapsulation arrangements. Therm. Sci. **23**, 823–833 (2019)
127. P. Bonnet, D. Sireude, B. Garnier, O. Chauvet, Thermal properties and percolation in carbon nanotube-polymer composites. Appl. Phys. Lett. **91**, 201910 (2007)
128. J. Wang, H. Xie, Z. Xin, Thermal properties of heat storage composites containing multiwalled carbon nanotubes. J. Appl. Phys. **104**, 113537 (2008)
129. S. Shaikh, K. Lafdi, K. Hallinan, Carbon nanoadditives to enhance latent energy storage of phase change materials. J. Appl. Phys. **103**, 94302 (2008)
130. S.L. Tariq, H.M. Ali, M.A. Akram, Thermal applications of hybrid phase change materials (HPCMs)-a critical review. Therm. Sci. **112** (2019)
131. Y. Zhang, B. Tang, L. Wang, R. Lu, D. Zhao, S. Zhang, Novel hybrid form-stable polyether phase change materials with good fire resistance. Energy Storage Mater. **6**, 46–52 (2017)
132. D. Su, Y. Jia, G. Alva, F. Tang, G. Fang, Preparation and thermal properties of n–octadecane/stearic acid eutectic mixtures with hexagonal boron nitride as phase change materials for thermal energy storage. Energy Build. **131**, 35–41 (2016)
133. X. Zhang, R. Wen, C. Tang, B. Wu, Z. Huang, X. Min, Y. Huang, Y. Liu, M. Fang, X. Wu, Thermal conductivity enhancement of polyethylene glycol/expanded perlite with carbon layer for heat storage application. Energy Build. **130**, 113–121 (2016)
134. H. Hong, Y. Pan, H. Sun, Z. Zhu, C. Ma, B. Wang, W. Liang, B. Yang, A. Li, Superwetting polypropylene aerogel supported form-stable phase change materials with extremely high

organics loading and enhanced thermal conductivity. Sol. Energy Mater. Sol. Cells. **174**, 307–313 (2018)
135. A. Karaipekli, A. Biçer, A. Sarı, V.V. Tyagi, Thermal characteristics of expanded perlite/paraffin composite phase change material with enhanced thermal conductivity using carbon nanotubes. Energy Convers. Manag. **134**, 373–381 (2017)
136. A. Sarı, Thermal energy storage characteristics of bentonite-based composite PCMs with enhanced thermal conductivity as novel thermal storage building materials. Energy Convers. Manag. **117**, 132–141 (2016)
137. J. Yang, L.-S. Tang, R.-Y. Bao, L. Bai, Z.-Y. Liu, B.-H. Xie, M.-B. Yang, W. Yang, Hybrid network structure of boron nitride and graphene oxide in shape-stabilized composite phase change materials with enhanced thermal conductivity and light-to-electric energy conversion capability. Sol. Energy Mater. Sol. Cells. **174**, 56–64 (2018)
138. M.A. Bashir, H.M. Ali, K.P. Amber, M.W. Bashir, H. Ali, S. Imran, M.S. Kamran, Performance investigation of photovoltaic modules by back surface water cooling. Therm. Sci. **22** (2018)
139. J. Yang, G.-Q. Qi, Y. Liu, R.-Y. Bao, Z.-Y. Liu, W. Yang, B.-H. Xie, M.-B. Yang, Hybrid graphene aerogels/phase change material composites: Thermal conductivity, shape-stabilization and light-to-thermal energy storage. Carbon N. Y. **100**, 693–702 (2016)
140. Y. Seki, Ş İnce, M.A. Ezan, A. Turgut, A. Erek, Graphite nanoplates loading into eutectic mixture of Adipic acid and Sebacic acid as phase change material. Sol. Energy Mater. Sol. Cells. **140**, 457–463 (2015)
141. S. Karaman, A. Karaipekli, A. Sarı, A. Bicer, Polyethylene glycol (PEG)/diatomite composite as a novel form-stable phase change material for thermal energy storage. Sol. Energy Mater. Sol. Cells. **95**, 1647–1653 (2011)
142. F. Tang, D. Su, Y. Tang, G. Fang, Synthesis and thermal properties of fatty acid eutectics and diatomite composites as shape-stabilized phase change materials with enhanced thermal conductivity. Sol. Energy Mater. Sol. Cells. **141**, 218–224 (2015)
143. M. Naguib, M. Kurtoglu, V. Presser, J. Lu, J. Niu, M. Heon, L. Hultman, Y. Gogotsi, M.W. Barsoum, Two-dimensional nanocrystals produced by exfoliation of Ti3AlC2. Adv. Mater. **23**, 4248–4253 (2011)
144. M. Naguib, V.N. Mochalin, M.W. Barsoum, Y. Gogotsi, Two-dimensional materials: 25th anniversary article: MXenes: a new family of two-dimensional materials (Adv. Mater. 7/2014). Adv. Mater. **26**, 982 (2014)
145. B. Anasori, M. Dahlqvist, J. Halim, E.J. Moon, J. Lu, B.C. Hosler, E.N. Caspi, S.J. May, L. Hultman, P. Eklund, Experimental and theoretical characterization of ordered MAX phases Mo2TiAlC2 and Mo2Ti2AlC3. J. Appl. Phys. **118**, 94304 (2015)
146. J.-C. Lei, X. Zhang, Z. Zhou, Recent advances in MXene: preparation, properties, and applications. Front. Phys. **10**, 276–286 (2015)
147. A. Szuplewska, D. Kulpińska, A. Dybko, M. Chudy, A.M. Jastrzębska, A. Olszyna, Z. Brzózka, Future applications of MXenes in biotechnology, nanomedicine, and sensors. Trends Biotechnol. (2019)
148. Y.-J. Zhang, J.-H. Lan, L. Wang, Q.-Y. Wu, C.-Z. Wang, T. Bo, Z.-F. Chai, W.-Q. Shi, Adsorption of uranyl species on hydroxylated titanium carbide nanosheet: A first-principles study. J. Hazard. Mater. **308**, 402–410 (2016)
149. R.M. Ronchi, J.T. Arantes, S.F. Santos, *Synthesis, Structure, Properties and Applications of MXenes: Current Status and Perspectives* (Ceram, Int, 2019).
150. B. Anasori, M.R. Lukatskaya, Y. Gogotsi, 2D metal carbides and nitrides (MXenes) for energy storage. Nat. Rev. Mater. **2**, 1–17 (2017)
151. aosterland, *Supporting Information*, Unpublished. 1803336, 1–12 (2012). https://doi.org/10.1002/adma.
152. C. Wang, H. Xie, S. Chen, B. Ge, D. Liu, C. Wu, W. Xu, W. Chu, G. Babu, P.M. Ajayan, Atomic cobalt covalently engineered interlayers for superior lithium-ion storage. Adv. Mater. **30**, 1802525 (2018)
153. X. Yin, C. Liang, Y. Feng, H. Zhang, Y. Wang, Y. Li, Research progress on synthetic scaffold in metabolic engineering—a review, Sheng Wu Gong Cheng Xue Bao= Chinese. J. Biotechnol. **35**, 363–374 (2019)

154. K. Rasool, R.P. Pandey, P.A. Rasheed, S. Buczek, Y. Gogotsi, K.A. Mahmoud, Water treatment and environmental remediation applications of two-dimensional metal carbides (MXenes). Mater. Today (2019)
155. K. Hantanasirisakul, M. Alhabeb, A. Lipatov, K. Maleski, B. Anasori, P. Salles, C. Ieosakulrat, P. Pakawatpanurut, A. Sinitskii, S.J. May, Effects of synthesis and processing on optoelectronic properties of Titanium Carbonitride MXene. Chem. Mater. **31**, 2941–2951 (2019)
156. K. Huang, Z. Li, J. Lin, G. Han, P. Huang, Two-dimensional transition metal carbides and nitrides (MXenes) for biomedical applications. Chem. Soc. Rev. **47**, 5109–5124 (2018)
157. L. Verger, C. Xu, V. Natu, H.-M. Cheng, W. Ren, M.W. Barsoum, Overview of the synthesis of MXenes and other ultrathin 2D transition metal carbides and nitrides. Curr. Opin. Solid State Mater. Sci. (2019)
158. G. Yang, B. Wang, H. Cheng, Z. Mao, H. Xu, Y. Zhong, X. Feng, J. Yu, X. Sui, Cellulosic scaffolds doped with boron nitride nanosheets for shape-stabilized phase change composites with enhanced thermal conductivity. Int. J. Biol. Macromol. **148**, 627–634 (2020)

Chapter 4
Thermophysical Properties of Advanced Energy Storage Materials

4.1 Summary

The various thermophysical properties of advanced energy storage materials, but not limited to, are thermal conductivity, latent heat capacity, density, phase change temperature and duration. These properties will be discussed in detail in this chapter.

4.2 Thermophysical Properties

4.2.1 Thermal Conductivity

The TC increment is directly proportional to enhancement in temperature. The inclusion of nano-enhanced particles boosts the TC of PCM, which results in increment in high value of "h". The value of increment in TC of PCM depends upon various factors like:

- Particles type and its concentration
- Shape and size of particle
- Temperature
- Type of surfactant (Fig. 4.1).

The value of TC of NPs is normally higher than PCM, which results in increment in the rate of heat transfer. The value of TC of MWCNTs is 3000 W/mK [1] while single-layer graphene has about 5300 W/mK [2]. The NPs amount must be added sufficient, which facilitates the heat transfer in PCM [3]. Zhang et al. [4] studied the properties of composite PCM in their experiments. The TC of erythritol having high aspect ratio of carbon nanofiber was high in comparison with carbon nanofiber of small aspect ratio.

H. M. Ali et al., *Thermal Energy Storage*,
https://doi.org/10.1007/978-981-16-1131-5_4

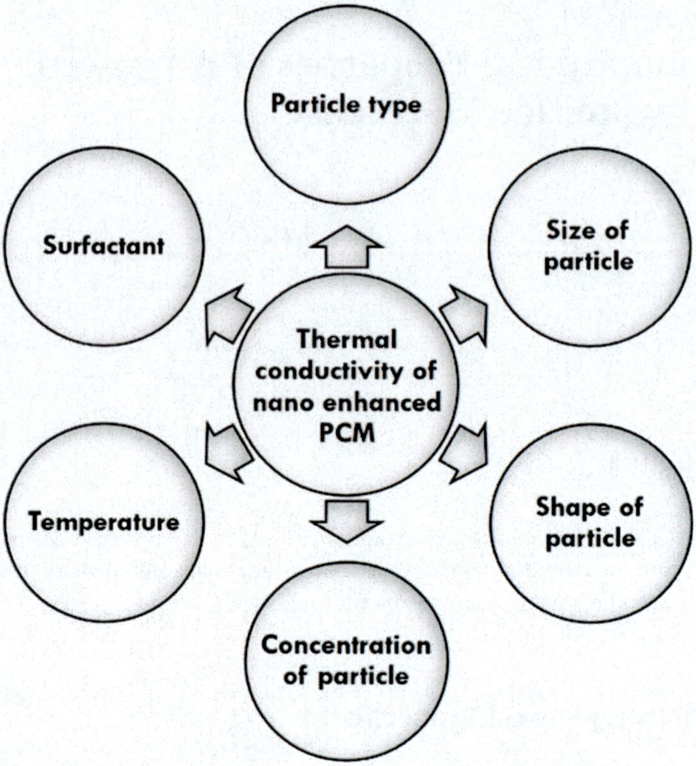

Fig. 4.1 Various factors enhancing TC of nano-enhanced PCM

The nano-enhanced PCM having high thermal conductive nanoparticles does not assure high TC rather various factors including particle homogeneous distribution, particle surface modification and surfactant addition must be considered. The surfactants are long organic molecules that alter particle surface properties along particle clustering [5]. But high amount of surfactants decreases the TC of base PCM. Gan et al. [6] study illustrated that the addition of surfactant decreased the TC of TiO_2/water and the same is the case with PCM. Different researchers concluded to a point that TC can also be enhanced by increasing the nanoparticle contents [7–13]. The researchers studied revealed that TC increased linearly and solid as well as liquid form of nano-enhanced PCM showed high value of TC. Chen et al. [14] studied recently the behavior of MWCNTs and steric acid on TC. Saeed et al. [15] investigated TC in solid and liquid phases of eutectic PCM with graphene platelets and results showed increment in TC in both phase at optimum concentration of NPs.

The TC of nano-enhanced PCM in liquid state is lower than solid phase. Wang et al. [16] studied MWCNTs with paraffin wax. The TC enhanced near melting point in solid state while it is decreased as temperature increased from melting point. The shape of nanoparticles also depends on aspect ratio of nanoparticles and it can be produced like sphere, cubic and rod-like cylindrical but cubic shape NPs are costly and complex [17]. So, more work to be needed to study the size and shape of nanoparticles that positively affect the TC of nano-enhanced PCM.

Thermal conductivity of encapsulated PCM is the major parameter for determining the efficiency of thermal energy storage capacity. The thermal conductivity of composite sphere single capsule can be represented as [18–20]

$$\frac{1}{k_p d_p} = \frac{1}{k_c d_c} + \frac{d_p - d_c}{k_s d_p d_c} \tag{4.1}$$

where k_p, k_c, k_s are the TC of micro/nanocapsule, core material and shell material respectively. While d_p and d_c are the diameters of micro/nanocapsule and core material respectively. Various instruments but not limited to, are used to measure TC of encapsulated PCM that includes TC3020 thermal conductivity meter [21], TC3000 thermal conductivity meter [22], EKO HC-110 thermal conductivity meter [23], laser flash apparatus [24]. The increment in TC of encapsulated PCM mainly depends upon shell material and organic coating materials exhibit lower TC while inorganic shell materials indicate higher value of TC. But on the other side, latent heat also decreases and thus there should be balanced while making encapsulated PCM.

4.2.2 *Latent Heat Capacity and Density*

Latent heat can be classified as latent heat of fusion and latent heat of vaporization. During sensible heating, the material temperature increases only and no phase change occurs. Most of the researchers [25–27] are concerned about latent heat along TC and concluded the following remarks regarding nano-enhanced PCM:

- The latent heat enhanced by modification of interaction between PCM and NPs
- The large quantity of NPs will degrade the latent heat of the composite.

Shaikh et al. [28] investigated SWCNTs, MWCNTs and carbon nanofiller-based nano-enhanced paraffin wax. The latent heat was increased as the concentration enhanced up to 1 vol%. Wang et al. [27] studied the behavior of paraffin wax by the addition of TiO_2 nanoparticles. The latent heat was increased up to 0.7 wt% of NPs. The latent heat started to decrease as the concentration further increased. A number of studies also found in which authors indicated that latent heat of PCM decreased with the addition of nanoparticles [29, 30]. Wang et al. [16] investigated latent heat effect on paraffin wax having MWCNT nanoparticles. The latent heat was decreased by 0.9% when MWCNTs added up to 2 wt%. Harikrishnan et al. [7] study demonstrated that property values of myristic acid having 1 wt% SiO_2 were decreased in comparison with single myristic acid. Lin and Al-Kayiem [29] revealed that latent heat was decreased by 14.6% when 2 wt% of copper addition was done in paraffin wax. The latent heat of solidification was also decreased by 13.3% with the same NPs concentration and the addition of hybrid NPs (copper-titania) also showed similar effects [30].

The density of composite PCM is generally increased by an increment of NPs concentration. The PCM having large density has high capability to transfer heat during phase change process [31, 32] but the research on this area is very limited. Warzoha et al. [13] found that density was enhanced up to 1.48% with paraffin wax having 8 wt% of graphite nanofibers compared to pure paraffin wax. Saeed et al. [15] carried out experiments and concluded that density of eutectic PCM was increased having graphene NPs. Similar results were seen with paraffin wax having graphene, MWCNTs, Al, TiO_2 NPs with 20 wt% [33]. The summary of various studies related to latent heat capacity is represented in Table 4.1.

4.2.3 *Phase Change Temperature and Duration*

The cooling is an important property of PCM, which influences thermal energy storage capacity [36]. When the effect of sub cooling is large then PCM will not be fully able to release heat properly. Latent heat releases when temperature will be lowered than solidification temperature. The inclusion of NPs into PCM may reduce this problem and NPs act as nucleation agent [37, 38].

Kumar and Kalaiselvam [12] investigated crystal growth with nuclei production that depends on nucleation effects. They observed the formation of ice crystals at the center although NPs were present at sides. As the TC of composite PCM is high compared to pure PCM, this will aid in energy utilizing and storing process. Harikrishnan et al. [7] study revealed that there was 30.69 and 32.32% time reduction in melting and solidification process when 1 wt% of SiO_2 NPs was added in myristic acid in comparison with pure myristic acid. The same trend was also observed by Warzoha and Fleischer [33] when paraffin wax was along graphene, MWCNTs, aluminum and TiO_2. Wang et al. [16] study illustrated that melting temperature was decreased from 53 to 52 °C by 2 wt% of MWCNT addition in PW. Lin and Al-Kayiem [29] deducted that there was reduction of melting temperature of 4.3% and solidification temperature increased up to 2.3% at 2 wt% of copper in paraffin wax. Harikrishnan et al. [7] conducted experiments on myristic acid having 1 wt% of SiO_2 and results indicated that melting and solidification time was increased about 1.13 and 1.30%, respectively. Rufuss et al. [35] found that there was a higher decrement in melting and solidification temperature range from 5.08 to 9.44% when paraffin wax was tested with TiO_2, CuO and graphene oxide NPs.

The various thermophysical properties of advanced energy storage materials are highlighted in this chapter. These properties include thermal conductivity, latent heat capacity as well as density and phase change temperature and duration. The selection of advanced energy storage is vital as the properties of base PCM largely depend on adding material on it. Although TC increased by using advanced energy storage material, but density and viscosity are also increased along this. There is also certain

Table. 4.1 Various studies done by different researchers on latent heat capacity

References	PCM	Nanoparticles	Concentration of NPs (wt%)	Latent heat capacity
[34]	Paraffin wax, Soy wax	Carbon nanofiber and tubes	1, 2, 5, 10	The change in latent heat capacity was not dominant after even addition of NPs
[7]	Myristic acid	SiO_2	0.2, 0.5, 0.8, 1.0	The addition of 1 wt% NPs slightly decrease the melting and solidification latent heat and it can be neglected
[8]	Paraffin wax, Micro-crystalline wax	α-alumina	0.5, 1.0, 2.0	Addition of NPs into both base fluids enhanced the latent heat capacity
[9]	Erythritol	Expanded graphite and carbon nanotubes	1, 3, 5, 7	The latent heat of the composite decreased with the increasing of NPs
[10]	Organic easter	Silver	0.1, 0.5, 1.0, 2.0, 5.0	The freezing and melting latent heat were decreased 7.3 and 8.2%, respectively, at 5 wt% of NPs in PCM
[12]	Palmitic acid	CuO	0.3, 0.5, 0.8 mass%	The latent heat for freezing changed from 156.8 to 148 kJ/kg as mass% of NPs altered from 0 to 0.8
[13]	Paraffin	Graphite nanofibers	2, 4, 6, 8	The latent heat was decreased with increment in NP concentration. The latent heat was 271.6 kJ/kg for paraffin while it was 242.7 kJ/kg for composite
[14]	Steric acid	MWCNTs	–	The latent heat of melting and freezing decreased with addition of MCNTs

(continued)

Table. 4.1 (continued)

References	PCM	Nanoparticles	Concentration of NPs (wt%)	Latent heat capacity
[16]	Paraffin wax	MWCNTs	0.2, 0.5, 1.0, 2.0	The latent heat for meting 165.3 kJ/kg in case of pure paraffin while it was 163.8 kJ/kg for composite at 2 wt% of MWCNTs
[26]	PW	CNFs, graphite, graphene	2.5, 5.0, 7.5, 10	The addition of NPs at low concentration enhanced the latent heat while it was not beneficial for high concentration
[27]	Paraffin wax	TiO_2	Up to 7	At concentration of 0.7 wt%, the latent heat of melting and freezing was increased but increasing 0.7 wt% concentration it decreased
[30]	Methyl cinnamate	Copper titania	0.05–0.5	The latent heat for freezing and melting decreased with increasing concentration of hybrid NPs
[35]	Paraffin wax	CuO, graphene Oxide, TiO_2	0.3	The increment or decrement of latent heat largely depends on type of NPs

limit too, if the temperature further increases then TC will decrease. So, this aspect should be kept in mind while selection of advanced energy storage material. The thermophysical properties are measured by different researchers but still there is no standard method to measure any specific property and researchers should consider this aspect in their future research. So, more work is to be needed in this domain to improve the performance and to measure various properties of advanced energy storage materials.

References

1. P. Kim, L. Shi, A. Majumdar, P.L. McEuen, Thermal transport measurements of individual multiwalled nanotubes. Phys. Rev. Lett. **87**, 215502 (2001)
2. A.A. Balandin, S. Ghosh, W. Bao, I. Calizo, D. Teweldebrhan, F. Miao, C.N. Lau, Superior thermal conductivity of single-layer graphene. Nano Lett. **8**, 902–907 (2008)
3. J.L. Zeng, Z. Cao, D.W. Yang, F. Xu, L.X. Sun, X.F. Zhang, L. Zhang, Effects of MWNTs on phase change enthalpy and thermal conductivity of a solid-liquid organic PCM. J. Therm. Anal. Calorim. **95**, 507–512 (2009)
4. Q. Zhang, Z. Luo, Q. Guo, G. Wu, Preparation and thermal properties of short carbon fibers/erythritol phase change materials. Energy Convers. Manag. **136**, 220–228 (2017)
5. Z. Mingzheng, X. Guodong, L. Jian, C. Lei, Z. Lijun, Analysis of factors influencing thermal conductivity and viscosity in different kinds of surfactant solutions. Exp. Therm. Fluid Sci. **36**, 22–29 (2012)
6. Y.Y. Gan, H.C. Ong, T.C. Ling, N.W.M. Zulkifli, C.-T. Wang, Y.-C. Yang, Thermal conductivity optimization and entropy generation analysis of titanium dioxide nanofluid in evacuated tube solar collector. Appl. Therm. Eng. **145**, 155–164 (2018)
7. S. Harikrishnan, S.I. Hussain, A. Devaraju, P. Sivasamy, S. Kalaiselvam, Improved performance of a newly prepared nano-enhanced phase change material for solar energy storage. J. Mech. Sci. Technol. **31**, 4903–4910 (2017)
8. N.H. Mohamed, F.S. Soliman, H. El Maghraby, Y.M. Moustfa, Thermal conductivity enhancement of treated petroleum waxes, as phase change material, by α nano alumina: energy storage. Renew. Sustain. Energy Rev. **70**, 1052–1058 (2017)
9. S. Guo, Q. Liu, J. Zhao, G. Jin, X. Wang, Z. Lang, W. He, Z. Gong, Evaluation and comparison of erythritol-based composites with addition of expanded graphite and carbon nanotubes. Appl. Energy **205**, 703–709 (2017)
10. R. Parameshwaran, R. Jayavel, S. Kalaiselvam, Study on thermal properties of organic ester phase-change material embedded with silver nanoparticles. J. Therm. Anal. Calorim. **114**, 845–858 (2013)
11. A. Elgafy, K. Lafdi, Effect of carbon nanofiber additives on thermal behavior of phase change materials. Carbon N. Y. **43**, 3067–3074 (2005)
12. K.R.S. Kumar, S. Kalaiselvam, Experimental investigations on the thermophysical properties of CuO-palmitic acid phase change material for heating applications. J. Therm. Anal. Calorim. **129**, 1647–1657 (2017)
13. R.J. Warzoha, A. Rao, R. Weigand, A.S. Fleischer, Experimental characterization of the thermal diffusivity of paraffin phase change material embedded with herringbone style graphite nanofibers, in *ASME 2012 Heat Transfer Summer Conferences Collocated with ASME 2012 Fluids Engineering Division Summer Meet. ASME 2012 10th International Conferences Nanochannels, Microchannels, Minichannels, American Society of Mechanical Engineers*, pp. 307–315 (2012)
14. Y. Chen, Q. Zhang, X. Wen, H. Yin, J. Liu, A novel CNT encapsulated phase change material with enhanced thermal conductivity and photo-thermal conversion performance. Sol. Energy Mater. Sol. Cells. **184**, 82–90 (2018)
15. R.M. Saeed, J.P. Schlegel, C. Castano, R. Sawafta, Preparation and enhanced thermal performance of novel (solid to gel) form-stable eutectic PCM modified by nano-graphene platelets. J. Energy Storage. **15**, 91–102 (2018)
16. J. Wang, H. Xie, Z. Xin, Thermal properties of paraffin based composites containing multi-walled carbon nanotubes. Thermochim. Acta. **488**, 39–42 (2009)
17. P.B. Maheshwary, C.C. Handa, K.R. Nemade, A comprehensive study of effect of concentration, particle size and particle shape on thermal conductivity of titania/water based nanofluid. Appl. Therm. Eng. **119**, 79–88 (2017)
18. M.I. Hasan, Numerical investigation of counter flow microchannel heat exchanger with MEPCM suspension. Appl. Therm. Eng. **31**, 1068–1075 (2011)

19. L. Liu, G. Alva, Y. Jia, X. Huang, G. Fang, Dynamic thermal characteristics analysis of microencapsulated phase change suspensions flowing through rectangular mini-channels for thermal energy storage. Energy Build. **134**, 37–51 (2017)
20. M. Jurkowska, I. Szczygieł, Review on properties of microencapsulated phase change materials slurries (mPCMS). Appl. Therm. Eng. **98**, 365–373 (2016)
21. X. Huang, Y. Lin, G. Fang, Thermal properties of polyvinyl butyral/graphene composites as encapsulation materials for solar cells. Sol. Energy **161**, 187–193 (2018)
22. K. Dutkowski, J.J. Fiuk, Experimental investigation on influence of microcapsules with PCM on propylene glycol rheological properties, in *E3S Web of Conferences*, p. 2005 (2018)
23. S. Yu, X. Wang, D. Wu, Microencapsulation of n-octadecane phase change material with calcium carbonate shell for enhancement of thermal conductivity and serving durability: synthesis, microstructure, and performance evaluation. Appl. Energy **114**, 632–643 (2014)
24. L. Chai, X. Wang, D. Wu, Development of bifunctional microencapsulated phase change materials with crystalline titanium dioxide shell for latent-heat storage and photocatalytic effectiveness. Appl. Energy. **138**, 661–674 (2015)
25. S. Sami, N. Etesami, Improving thermal characteristics and stability of phase change material containing TiO_2 nanoparticles after thermal cycles for energy storage. Appl. Therm. Eng. **124**, 346–352 (2017)
26. F. Bahiraei, A. Fartaj, G.-A. Nazri, Experimental and numerical investigation on the performance of carbon-based nanoenhanced phase change materials for thermal management applications. Energy Convers. Manag. **153**, 115–128 (2017)
27. J. Wang, H. Xie, Z. Guo, L. Guan, Y. Li, Improved thermal properties of paraffin wax by the addition of TiO_2 nanoparticles. Appl. Therm. Eng. **73**, 1541–1547 (2014)
28. S. Shaikh, K. Lafdi, K. Hallinan, Carbon nanoadditives to enhance latent energy storage of phase change materials. J. Appl. Phys. **103**, 94302 (2008)
29. S.C. Lin, H.H. Al-Kayiem, Evaluation of copper nanoparticles—paraffin wax compositions for solar thermal energy storage. Sol. Energy. **132**, 267–278 (2016)
30. K.R.S. Kumar, R. Parameshwaran, S. Kalaiselvam, Preparation and characterization of hybrid nanocomposite embedded organic methyl ester as phase change material. Sol. Energy Mater. Sol. Cells. **171**, 148–160 (2017)
31. H. Schmidt, H. Wolf, E. Hassel, A method to measure the density of seawater accurately to the level of 10–6. Metrologia **53**, 770 (2016)
32. I. Mjallal, H. Farhat, M. Hammoud, S. Ali, I. Assi, Improving the cooling efficiency of heat sinks through the use of different types of phase change materials. Technologies. **6**, 5 (2018)
33. R.J. Warzoha, A.S. Fleischer, Improved heat recovery from paraffin-based phase change materials due to the presence of percolating graphene networks. Int. J. Heat Mass Transf. **79**, 314–323 (2014)
34. Y. Cui, C. Liu, S. Hu, X. Yu, The experimental exploration of carbon nanofiber and carbon nanotube additives on thermal behavior of phase change materials. Sol. Energy Mater. Sol. Cells. **95**, 1208–1212 (2011)
35. D.D.W. Rufuss, L. Suganthi, S. Iniyan, P.A. Davies, Effects of nanoparticle-enhanced phase change material (NPCM) on solar still productivity. J. Clean. Prod. **192**, 9–29 (2018)
36. A. Safari, R. Saidur, F.A. Sulaiman, Y. Xu, J. Dong, A review on supercooling of phase change materials in thermal energy storage systems. Renew. Sustain. Energy Rev. **70**, 905–919 (2017)
37. B.M.L.G. Ramirez, C. Glorieux, E.S.M. Martinez, J.J.A.F. Cuautle, Tuning of thermal properties of sodium acetate trihydrate by blending with polymer and silver nanoparticles. Appl. Therm. Eng. **62**, 838–844 (2014)
38. K.P. Venkitaraj, S. Suresh, B. Praveen, A. Venugopal, S.C. Nair, Pentaerythritol with alumina nano additives for thermal energy storage applications. J. Energy Storage. **13**, 359–377 (2017)

Chapter 5
Energy Storage Materials in Thermal Storage Applications

5.1 Summary

This chapter contains applications of advanced energy storage materials in a broad range that includes, but not limited, in buildings, solar energy, waste heat recovery, seawater desalination, electronic cooling and photovoltaic thermal systems. The major use of PCMs in TES for several applications is presented as PCM can reduce operation cost, capacity and size of these applications where TES is applied. PCM plays an important role in fulfilling and efficient along environmentally friendly usage of thermal energy in various areas PCMs are used in.

- TES for buildings
- Nuclear TES
- Air conditioning applications
- Solar Photovoltaic system
- Solar water heating system
- Biomedical applications
- Storage and transport applications
- Temperature sensitive products
- Electronics

5.2 Thermal Storage Applications

The energy storage materials have been widely used in various practical applications and researchers continuously studied the properties of these materials to broad the application area. The statistics of "Web of Science" illustrated that there are 43,103 results found till 2018 when searched with the keyword "energy storage materials". The search results clearly showed that a lot of work has been done on this domain. The energy storage materials have been extensively used in a wide range of applications as solar heating systems [1–4], building air conditioning [5–8], electronic cooling

H. M. Ali et al., *Thermal Energy Storage*,
https://doi.org/10.1007/978-981-16-1131-5_5

[9–12], waste heat recovery [13–15], textiles [16, 17] and a few more one. In recent years, direct solar steam generation (DSSG) method has been studied for clean water [18–25]. The advanced materials are used for water having high evaporation rate [26–34]. The advanced energy storage materials have also been extensively used in hybrid/photovoltaic thermal system.

5.2.1 Thermal Energy Storage in Building Applications

The phase change material contains certain properties for thermal energy storage in building applications. The properties included thermodynamic and economical characteristics [35, 36]. Some of the main characteristics are:

- The PCM must have low density, high specific heat and high thermal conductivity and low vapor pressure.
- A little or no supercooling and this property can be achieved by high rate of crystal growth or high nucleation rate.
- The PCM must be chemically stable and non-corrosive to building material.
- The PCM must be effective, show reliability and compatibility.
- Easily availability.

The working temperature of PCM must be considered when used in building applications. The PCM should be capable for working in desired temperature range, local climate conditions, location where it is used and phase transition temperature must be 18–30 °C temperature range, which can then be used in building applications [37, 38]. Kalnaes and Jelle [39] suggested three temperature range for use of PCM in buildings, which are:

- Up to 21 °C for cooling applications
- 22–28 °C for human comfort applications
- 29–60 °C for hot water applications.

There are several ways for assimilating PCM in buildings and it may be used in active or passive storage system or combination of both [40]. In case of passive storage system, the process proceeds due to difference of temperature between PCM and surrounding while in active systems, it occurred with the help of electrical appliances as fans, pumps (Fig. 5.1).

5.2.1.1 Passive System

The passive storage system gains special attention of interest and, in this case, the PCM can be added as separate component in building. The method of applying PCM in building should be such that heat can be utilized or stored if temperature of surrounding is low or high than melting temperature of PCM.

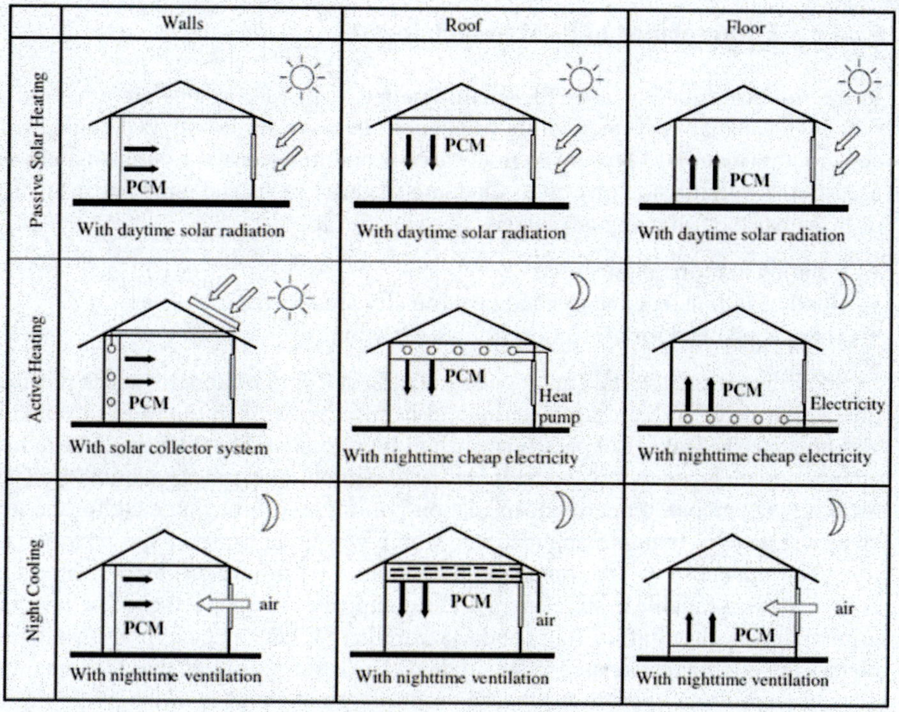

Fig. 5.1 The PCM applications in buildings [41]

5.2.1.2 Various Conventional Techniques

The direct incorporation technique has been investigated by various authors [42–45] and in this method, porous building material is dipped into hot melted PCM. The porous material is removed from liquid PCM and PCM remains in porous of building material [46].

The direct incorporation technique is most economical due to little additional process equipment's. Liquid or powdered PCM is mixed with building materials as gypsum or concrete. The energy is stored in gypsum wallboard by direct incorporation [47] of PCM in it. But these methods may leakage of PCM over long period of time.

The PCM can be macro-encapsulated before incorporation in which PCM is packed in tubes, pouches, spheres or panels [48]. Zhang et al. [49] developed and performed tests on highly crystalline paraffin wax and the results showed that wall reduced heat fluxes to about 38%.

Direct immersion or microencapsulation has also disadvantaged that they should be protected against destruction, difficult to integrate into structure, which makes it expensive and has poor heat transfer coefficient [50].

5.2.1.3 Encapsulation of PCM in Buildings

In micro-encapsulation of PCM, small amount of PCM particles are enclosed in thin, high weight of polymeric film, which must be compatible with PCM along with construction material. Gypsum is widely used in building construction and it is always used in the interior side of wall as cladding material [51]. The microencapsulation of PCM has the following advantages:

- Installed in more applications
- Good heat transfers due to good heat transfer area surface
- There is no need for protection from destruction.

Shossig et al. [52] studied microencapsulation of PW in wallboard, which prevents interaction between building material and paraffin wax. Such micro-encapsulation also prevents the leakage of paraffin from building element. The fluctuations in indoor temperature reduced to 2 °C compared to wall without microencapsulation of PCM. Su et al. [53] prepared micro-encapsulation-paraffin/gypsum composite for thermal energy storage in building applications, which was done using in-situ polymerization. The thickness of the composite was about 1.0 cm and results indicated that composite was capable to store energy. The composite also showed good behavior in minimizing electric traffic. Toppi and Mazzarella [54] developed gypsum wallboard containing microencapsulated PCM product. The results revealed that little quantity of microcapsule in gypsum wallboard produced great changes in the properties of gypsum.

Castellon et al. [55] investigated microencapsulated PCM in concrete building in Spain. The results indicated positive effects on excess temperature and changes were occurred inside the cubical. Gypsum ceiling panels impregnated with PCM for reducing indoor air temperature and cooling load in test building was done in Malaysia [56]. The numerical method was used to check the impact of PCM and results indicated that for maintaining room temperature, the building with PCM required less amount of energy for cooling energy. Scalat et al. [57] conducted experiments in gypsum boards with or without PCM and gypsum wallboard was impregnated with 25% PCM. When rooms were heated at 26 °C and cooled down to 12 °C then results showed that temperature of the room having equipped PCM decreased more slowly than room without PCM and results were also same for heating the room. Kunzik et al. [58] investigated PCM board mounted on walls or ceiling, which has a thickness of 5 mm and 60% microencapsulated PCM having melting temperature of 22 °C was used in it. The tests were conducted in temperature range varied from 15 to 30 °C and energy stored in PCM wall was twice when PCM board used with additional 8 cm thickness of concrete. Also, the temperature variation was lowered about 4.7 °C when used PCM in wall. The microencapsulation of PCM has also been investigated in the Europe climate for energy savings in building applications. Cabeza et al. [59] studied microencapsulated PCM for saving energy in walls of concrete of various homes. The results indicated that thermal inertia improved in walls having capsulated PCM compared to conventional concrete wall and inner temperature was also lowered as illustrated in Fig. 5.2.

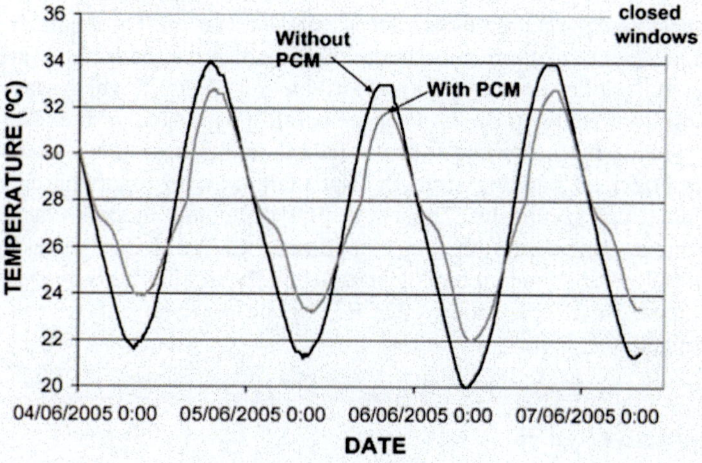

Fig. 5.2 Difference of temperature with and without PCM on south wall of house [59]

Arce et al. [60] developed encapsulated PCM in cubical wall of concrete. The results illustrated that there was reduction of peak temperature of 6% and comfort time enhanced up to 10%. Thiele et al. [61] made microencapsulated PCM-concrete composite used in building for energy saving. The effect of volume fraction of PCM, temperature of PCM, window temperature and outdoor temperature were under examination. The addition of capsulated PCM late thermal load in various climate conditions and thermal was improved. The unique kind of PCM is also used in recent years known as shape stabilized PCM (SSPCM) and has been under consideration by different authors [62–67]. The SSPCM contains paraffin and polyethylene of high density or other such material. Zhang et al. [68] undergone the testing of SSPCM samples as illustrated in Fig. 5.3 for determining thermophysical properties and space

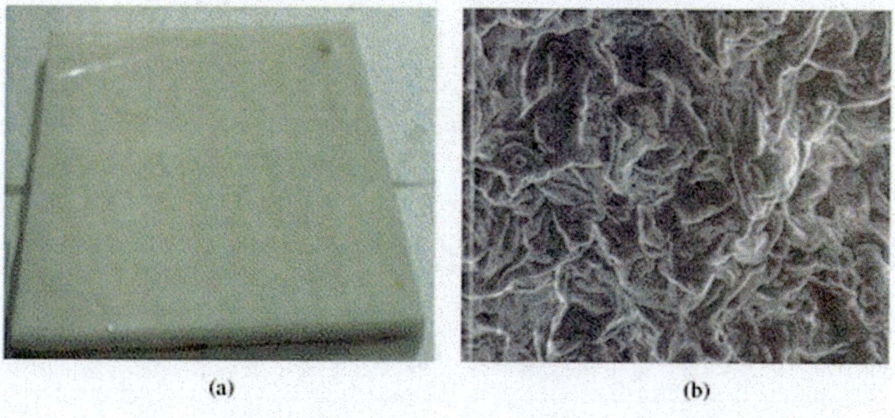

Fig. 5.3 **a** Picture of plate contained SSPCM, **b** SEM image of PCM HITACHI S-450 [68]

heating performed simulations as well as experimentation. The results indicated that SSPCM has potential application in buildings and should be studied further.

Lie et al. [66] developed gypsum board having SSPCM, which was synthesized by mixing water and gypsum and adding paraffin/expanded perlite. The strength of SSPCM was tested according to the standards and adsorption amount of paraffin has little effect on bending strength. But there was no report on energy storage properties of this composite. Zhou et al. [69] simulated thermal characteristics of middle direct gain room with shape stabilized PCM (SSPCM) plates as inner and investigated melting temperature, heat of fusion, board thickness of SSPCM thermal performance. The results showed that PCM plates have advantages in solar housed for direct gain.

5.2.1.4 Active System

The energy storage material with solar heat pump system, heat recovery system, HVAC system, floor heating system [39] are called active systems. Such systems may be gathered for peak load reduction achievement. The peak load during daytime put pressure on grid and consequently enhances the demand for cooling or heating loads by HVAC systems. Using PCM, the peak load demand may be divided during day resulted in reducing the highest peaks. Figure 5.4 illustrated that how load is shifted by using PCM.

Sun et al. [70] investigated PCMs for testing of peak load shifting as well as controlling and suggested that reduction of peak cooling load is from 10 to 57%. The peak reduction was found when PCM was compared lightweight insulated construction. Furthermore, the authors proposed that more proper load control plans can result in more cost savings.

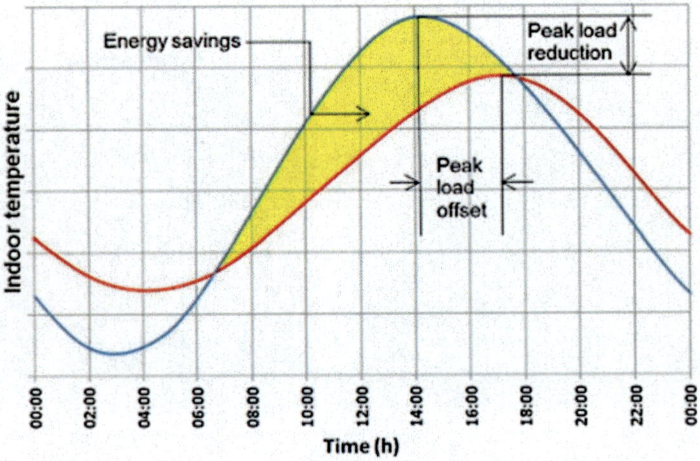

Fig. 5.4 Illustration of reduction of peak load [39]

5.2.2 Energy Storage Materials in Solar Energy Applications

For the conversion of solar energy into thermal energy, the devices that are mostly used are solar concentrators and collectors. The various researchers done work on various types of fluids to enhance the maximum efficiency of these systems and the energy storage materials have wide applications in solar energy. The under consideration of energy storage materials in solar energy has caught special attention in recent years. This is due to maximum gain of solar radiations by utilizing energy storage materials.

The energy storage materials can be used in parabolic trough collectors. Kumaresan et al. [71] examined the energy storage material in solar parabolic tough collector (PTC). The system consisted of parabolic trough collector, TES tank and pump with the aim of determining thermal efficiency of system. The proposals were given to minimize the heat losses and idea of using storage material minimized the heat losses. Molten salts are widely used in parabolic trough collectors and concentrated solar power plants due to its better thermal properties. These two technologies used two-tank system as thermal storage system and the concept was illustrated by solar thermal demo plant [72, 73] (Table 5.1).

Zipf et al. [74] investigated various PCM storages in 50 MW CSP plant along heat exchangers. A thermal storage can enhance the solar thermal electricity and yearly yield estimation was evaluated by using such heat exchanger. The cost analysis of this power plant was also done. This configuration resulted in achieving high outlet steam temperature and 24% of salt was superheated. A solar thermal power plant located in Shiraz, Iran was investigated by Mahfuz et al. [75]. The thermodynamics properties were considered under the inclusion of PCM storage. The results indicated that efficiency of exergy of system with PCM was higher than 30% of energy efficiency. Bhagat and Saha [76] studied rankine cycle based on solar power plant with encapsulated PCM (EPCM). The various properties were considered as charging temperature, size of storage system, EPCM shell diameter as well as porosity. The results indicated that increment of mass flow rate and charging temperature enhanced TC. The fluid temperature decreases with the decrement in EPCM shell diameter.

Table 5.1 Thermal energy storage by molten salts in demo plants [72, 73]

TES demo plant	Type of fluid for HT	Hot storage temperature (°C)	Hot tank volume (m^3)	Thermal capacity (MWh_{th})
Thermis (France)	Hitectm	450	310	40
CRTF (France)	Solar salt	566	53	7
Solar two (USA)	Solar salt	565	875	105
Archimede (Italy)	Solar salt	550	25	4

Lin and Al-Kayiem [77] studied solar thermal storage with nano PCM containing copper NPs in paraffin wax and it was observed that increasing the concentration of NPs enhanced the TC value of base PCM. It was also concluded that at 1% concentration of NP, the efficiency of paraffin wax increased by 1.7%. Singh et al. [78] enhanced the thermal performance of solar storage system by addition of copper oxide and alumina NPs in base PCM. The melting point of base PCM was decreased by addition of copper oxide NPs while increased with alumina NPs. The thermal energy storage was better in nano PCM of alumina-based PCM than copper oxide-based nano PCM, but the distribution of copper oxide-based nano PCM was more even after 50 cycles than alumina-based nano PCM. Galione et al. [79] investigated 50 MW plant by using multi-layered solid PCM (MLSPCM) for concentrated solar power (CSP). In MLSPCM, the PCM at top has high melting point and at base has low melting point as illustrated in Fig. 5.5. The MLSPCM exhibited great storage capacity and large production of utilization.

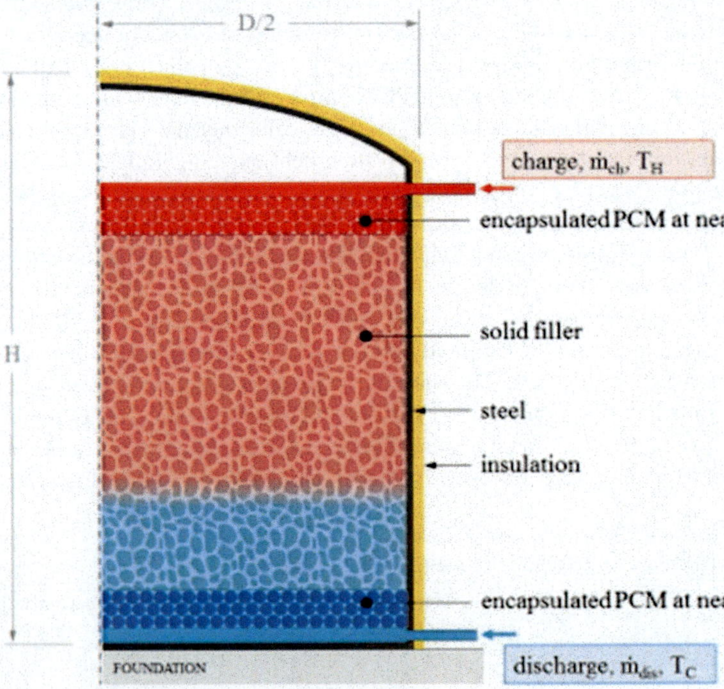

Fig. 5.5 Schematic of three-layer MLSPCM [79]

5.2.3 *Waste Heat Recovery Storage from Industrial Applications*

The thermal energy storage is the available possibility to solve problems related to industrial waste heat to gain better capacity factor. The typical capacity for PCM is 50–150 KWh/ton and storage period can be hourly or seasonal while the cost range 10–50 ϵ/KWh [80]. The thermal energy can be used on-site or off-site. The onsite is when waste heat source and heat demand on same site while off-site is vice versa as illustrated in Fig. 5.6.

Steinmann [81] suggested latent heat TES tank based on graphite and mixture of KMO_3 and $NaNO_3$ as PCM to recover waste heat from aerated concrete manufacturing industry. The PCM was obtained with saturated steam at 3 bar. Selvaraj et al. [82] carried out study on moulds having aluminum shots as TES to recover waste heat. The system consisted of moulds of sand having aluminum shots that surrounded cavity of mould. It was estimated that 6.4% waste heat can be collected and used in further process. Maruoka et al. [83] investigated TES contained balls of copper capsulated by film of nickel to store industrial waste heat from exhaust above 1600 °C. The stored heat was further processed in the chemical reaction for forming methanol and to accelerate the reaction. Zhang and Akiyama [84] analyzed sodium acetate trihydration salt as PCM for thermal energy storage to store industrial waste

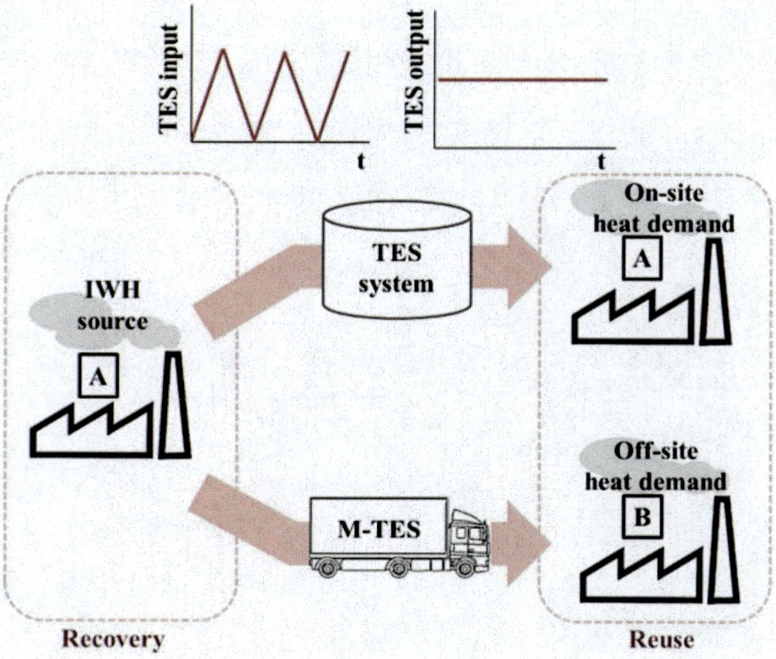

Fig. 5.6 Schematic for on-shore and off-shore for industrial waste heat [81]

heat. The authors concluded that exergy losses reduced significantly as temperature is controlled, however, saving was not calculated.

The industrial waste heat was also studied by various authors [85–87] to investigate the coupling of TES system to diesel engine. The experimental setup contained water cooled diesel engine, which was integrated into recovery heat exchanger connected to TES system. The TES system consisted of [85, 87] 48 cylindrical capsules having paraffin embedded in it and 55 kg of caster oil was used. The results showed that there was 15.2% reduction of fuel consumption in comparison with setup without TES. The same setup was also investigated by Prabu and Asokan [88] and Gopal et al. [89], which consisted of TES system. The vessel contained 40 spherical containers containing 100 kg paraffin PCM each in it. The water was used as heat transfer fluid and results indicated that approximately 7% of exhaust heat was recovered. Steinparzer et al. [87] investigated exhaust gas from electric arc furnace as industrial waste heat with PCM and molten salt as TES for reuse of heat represented in Fig. 5.7. The estimated power was 370 KWh/ton of liquid steel and results indicated that 24% of energy could be saved on-site. The reduction in energy consumption was 60–80 KWh/ton and CO_2 emission could be reduced 45 kg/ton of liquid steel. A test rig having 4500 kg molten salt was also modeled but published results were not found.

Kauranen et al. [90] investigated the effect of TES system to diesel engine having outdoor under zero conditions. The 4 kg of Climsel C70 having melting temperature of 70 °C was used to accumulate heat. The results of this system were better and cold start emissions were decreased when compared to original system. Shon et al. [91] suggested system to store waste heat from engine in tank having 4.2 kg of xylitol with 1300 kJ heat capacity. The stored heat was utilized in heating engine and results

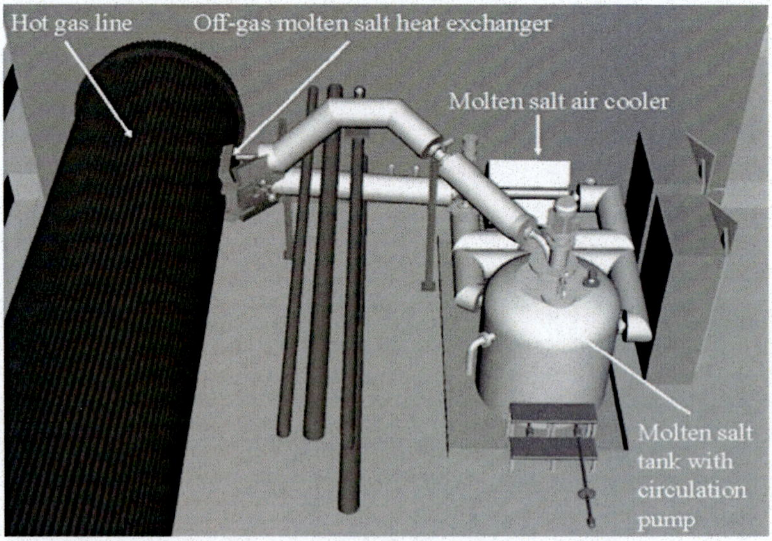

Fig. 5.7 Schematic diagram of the proposed pilot plant [87]

Table 5.2 Waste heat recovery studied by different authors

Industry sector	Waste heat source	Exhaust temperature (°C)	TES media	Savings	References
Basic metals	Steel works	430	Molten salt	60–80 KW/ton of energy saved	[87]
Non-metallic minerals	Cement production plant	177–816	Rock bed and salts	24 × 10^{13} Btu/year energy saved	[93]
	Concrete production plant	–	Eutectic mixture of KNO_3 and $NaNO_3$ with graphite	–	
	Concrete production plant	–	Reinforced concrete slabs and paraffin	45 % energy consumption saved and 15,000-ton CO_2.year	
Chemical products	Organic surfactant plant	110–160	Metal encapsulated PCM	Up to 56% energy saved	[94]
Basic metals	Metal casting	1200–1600	Molten slag	–	[95]

illustrated that warming time of engine was decreased up to 33.7%. Baldi et al. [92] designed and modeled TES system for shipping transportation. The TES system was 1000 m^3 cylindrical storage tank contained thermal oil and it was observed that 80% of fuel consumption was reduced from boiler. The various studies on waste heat recovery investigated by different authors are represented in Table 5.2.

5.2.4 *Energy Storage Materials in Seawater Desalination*

Energy storage materials including thermal energy storage materials are widely used in seawater desalination processes especially in solar still. These advanced materials have extraordinary properties, which absorb the maximum solar radiations and convert it to water evaporation. The various energy storage materials for desalination are represented in Fig. 5.8. However, only thermal energy storage materials will be discussed here for application point of view.

In steam power plants, waste heat can be get rid by cooling towers into surrounding air. This heat can be utilized in desalination plant. Elminshawy et al. [97] suggested desalination plant from industrial waste heat recovered from various plants in Saudi Arabia. A test rig was also constructed to store waste heat from steam turbine using

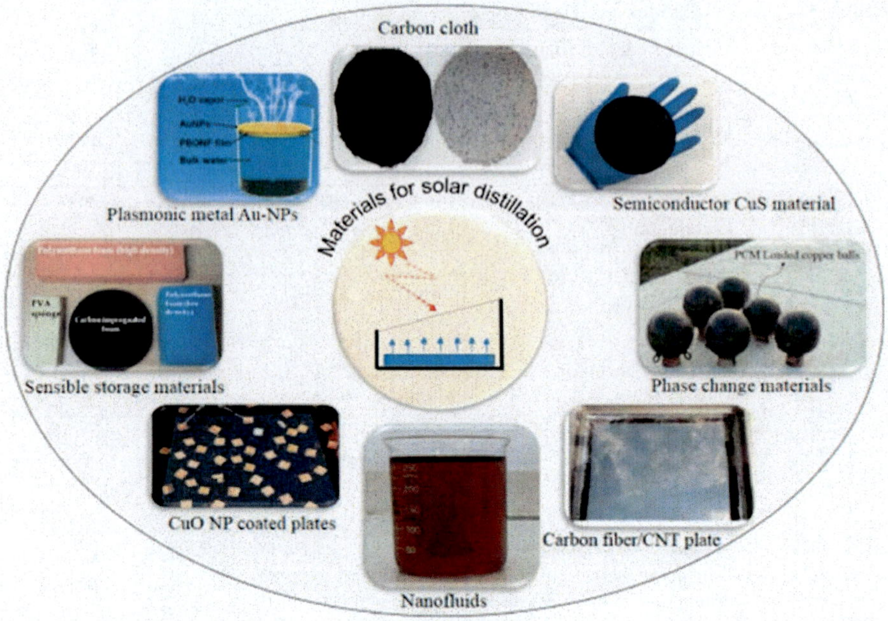

Fig. 5.8 Advance materials for desalination purposes [96]

water tank as TES system. The results showed that 47 MW steam turbine saved 1844 kg/h fuel consumption.

Gude [98] analyzed desalination plant with solar and waste heat as energy source as illustrated in Fig. 5.9, which resulted in better efficiency of plant. The molten salt, concrete, PCM were proposed to store waste heat. The hot water was suggested for solar still and other low-temperature applications.

Faegh and Shafii [99] conducted experiments by integrating solar still with heat pipe, evacuated tube collector and phase change material. The productivity was 6.5 kg/m^2 and 50 daily efficiency was achieved by this integration. Sharshir et al. [100] studied the solar still with paraffin wax and graphite NPs in water. The enhancement

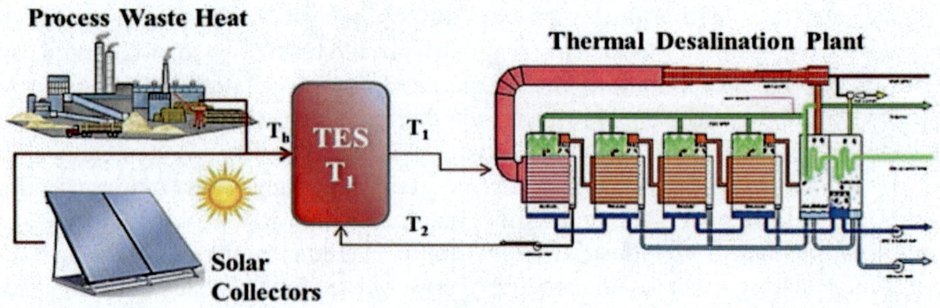

Fig. 5.9 Desalination plant coupled with solar collector and industrial waste heat [98]

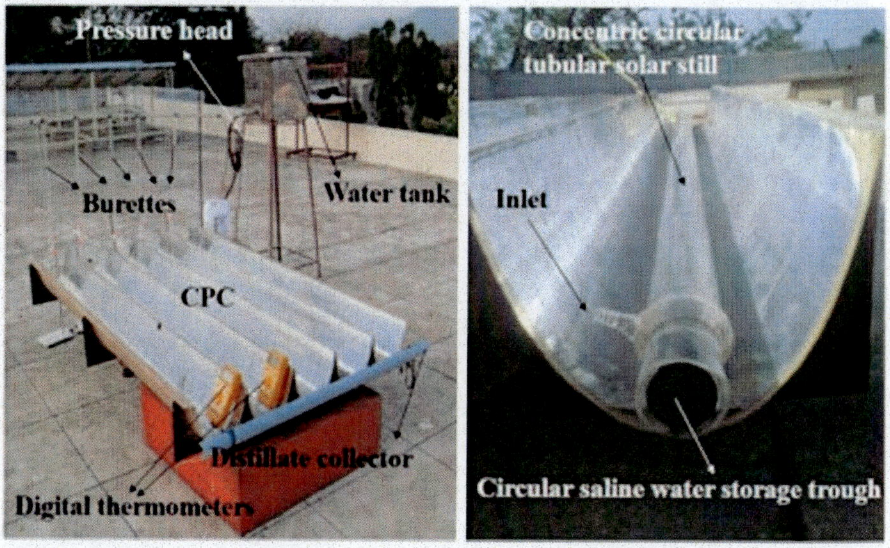

Fig. 5.10 Paraffin as latent heat storage in concentric circular tabular still [101]

in water productivity was up to 65% and this increment was due to latent heat storage materials in water. The authors also investigated film cooling method with solar still and achieved 73.8% improvement in water productivity.

Arunkumar and Kabeel [101] studied circular tubular still integrated with latent heat storage materials. The amount of water produced was 5.7 kg/m^2 and the experimental setup is represented in Fig. 5.10. The authors concluded that combined effect of latent heat storage and value of "k" improved the system performance by about 8% (Table 5.3).

5.2.5 Energy Storage Material in Cooling Devices

Thermal management technologies of handheld electronic devices have become significant area of research. Devices like smartphones, computers, tablets, personnel digital assistant, digital cameras etc. are based on more advanced electronic components, complex microcircuits, additional features and use high power during operation [108, 109]. Both the size and weight of devices have decreased with performance remarkably increased. However, the high-power levels consequently generate more heat and high temperature produced during the operation of these devices may lead to overheating, runaway, combustion, failure of critical components and also creating discomfort for users. This is because in many situations, the power usage may be high due to playing multiple functions at a time, for example, listening to the music while

Table 5.3 Advanced materials investigated by various authors for steam generation

Evaporator type	Nanoparticles size (nm)	Intensity of solar radiations $\left(\frac{KW}{m^2}\right)$	Rate of water evaporation $\left(\frac{kgm^{-2}}{h}\right)$	Efficiency (%)	References
Reduced graphene oxide/polyurethane	300–1500	10	0.9	65	[102]
Graphitized carbon black	400	10	–	69	[103]
SiO_2/Au nano shell	1000	0.0235	–	80	[104]
Paper-based Au NP film	17.4	4.5	–	57	[105]
N-doped porous graphene	100–300	1.0	–	54	[106]
Carbon foam/Graphite layer	–	10	–	85	[107]

playing game etc. Due to compact size, the area for heat dissipation is small. For this reason, very big and serious problem is faced by electronic industries [110–114].

The researchers all around the world attempted experimental and numerical investigations to obtain effective thermal management of these devices to solve this extra heat generation issues. Many techniques are applied to keep the surface temperature within safe and comfortable range, e.g. thermoelectric cooling, air cooling, novel PCM cooling, etc. [115]. In many old electronic devices, conventional active cooling methods were popular to keep cool these devices. Forced conventional techniques using macro fans to increase the heat transfer rate are effective, but they are not favorable as they make the system massive, noisy, costly and bulkier. They also consume high power and encounter more often maintenance and other problems. Due to these drawbacks, traditional cooling methods are considered unsuitable for the current devices [116].

The cooling of electronic devices plays a vital role in their functional operation like human mind and various techniques that include energy storage materials, PCM materials are used for cooling of electronic devices [117, 118]. These materials have gained practical benefits for thermal management of electronic devices. It is a great challenge for power industry to dissipate heat at rate of 300 W/cm^2 while maintaining the temperature 85 °C [119]. The cooling system integrated with energy storage materials enhanced its performance [120].

Krishna et al. [121] investigated heat pipe with non-enhanced PCMs for electronic cooling applications as schematic diagram is illustrated in Fig. 5.11. The PCMs store or release thermal energy according to input power. The water, Tricosane PCM and Trocosane with nano alumina particles were investigated as energy storage materials. The experimental results indicated that nano-enhanced PCM has high TC than pure

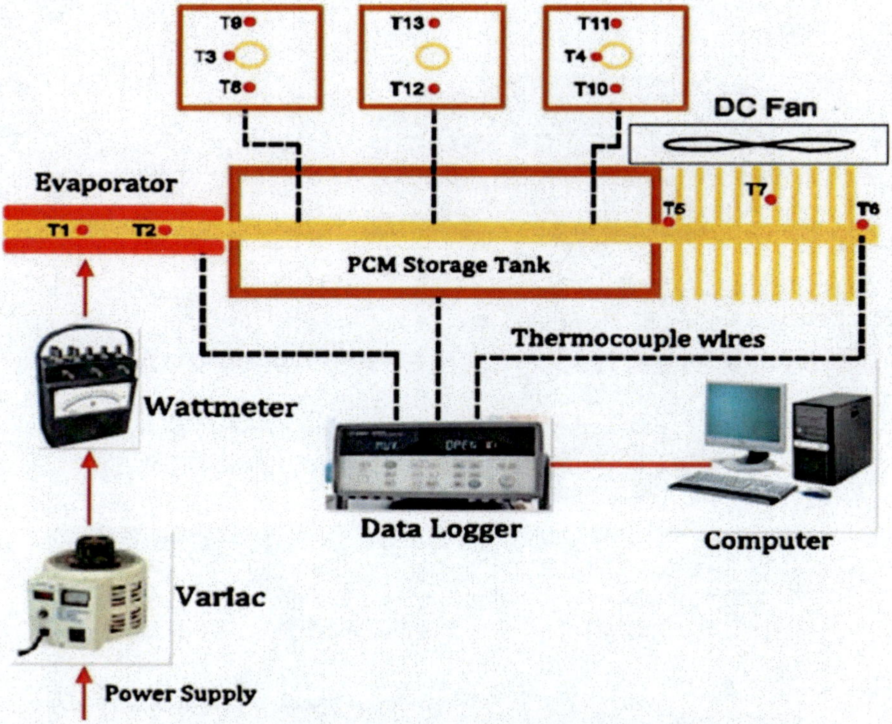

Fig. 5.11 Schematic diagram of the proposed experimental setup [121]

PCM and maximum enhancement was 32%. The heat pipe integrated with nano-enhanced PCM reduced the evaporator temperature by 25.75%.

Itani et al. [122] studied the cooling vest using PCMs and main aim was to find better position and number of PCM packets to be used to achieve optimum cooling effects. The results indicated that number of PCM packets required was 8, 18 and 20 when outdoor temperature was 28, 35, 45 °C respectively. Thus, various amounts of PCM packets are required according to environmental conditions. Yasufogolu et al. [123] integrated energy storage materials for domestic refrigerator. The four kinds of PCMs were used for testing on two refrigerators. The results indicated that opening and cooling time was optimized and 0.95 kg of PCM saved 9.4% energy.

Bakshipour et al. [124] used PCM in household refrigerator and results illustrated that power consumption was reduced and coefficient of performance (COP) was improved in comparison to without PCM. Chaiyat [125] integrated PCM with split air-conditioner to decrease the air temperature that enters the evaporation coil. The power consumption was reduced by 9% by using this technique with 4.12 years of pay back period. Said and Hassan [126] also used same type of approach and achieved high energy saving and COP. Alimohammadi et al. [127] used nano PCM for cooling of electronic chipset under natural and forced convection as heat sink section and experimental setup is shown in Fig. 5.12. The nano PCM contained 1

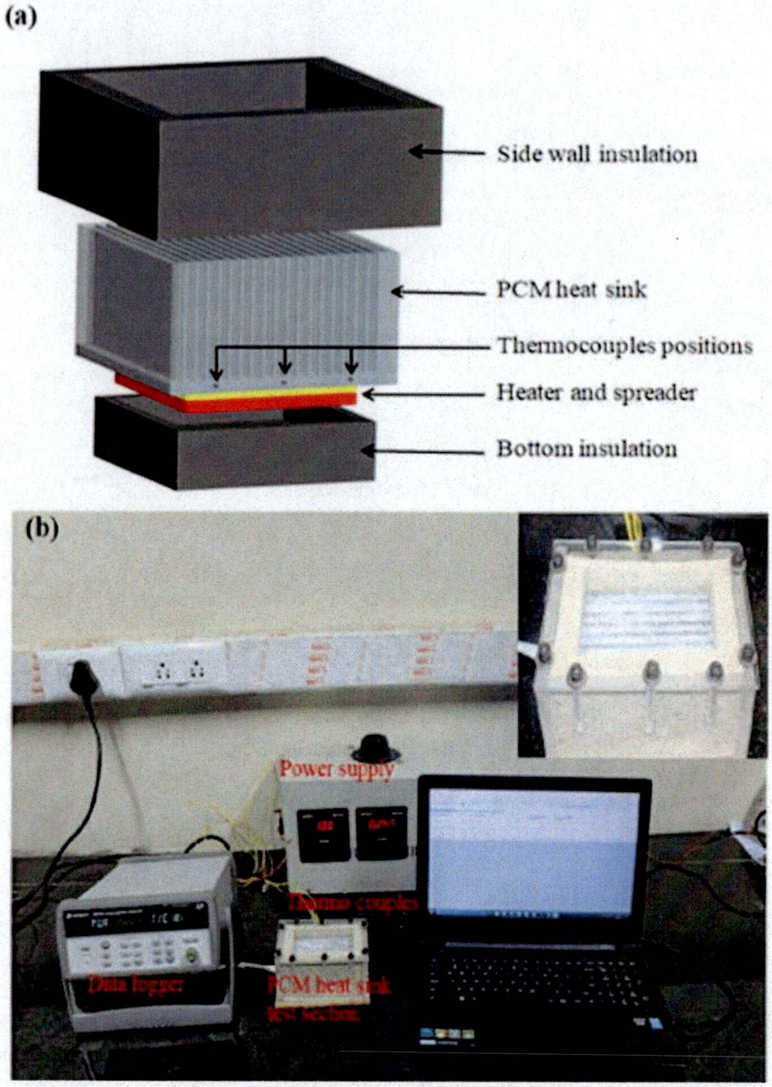

Fig. 5.12 **a** Schematic diagram of test section of heat sink, **b** Experimental setup [128]

wt% of iron oxide-based PCM of manganese nitrate and it was observed that nano PCM decreased that chip temperature by 14 and 10.5 °C by forced and natural convection respectively.

Praveen and Suresh [128] studies HT and thermal properties of heat sink using nano PCM and copper oxide NPs in Neopentyl glycol were used for nano PCM formation. The results indicated that decrement in TC was 1.16–4.08 times, reduction in energy storage time 14.1–33.8% by increasing NPs concentration from 0.5

to 3%. Haung et al. [129] enhanced the TC from 14.9 to 59.6 W/mK by adding graphite in base PCM in electronic board cooling. Farhzanehia et al. [130] investigated the performance of PCM and nano PCM in electronic chipsets applications. The results indicated that addition of MWCNTs in paraffin wax (nano PCM) decreased the cooling time by 6% compared to pure PCM. The maximum temperature was decreased, and operating temperature was also increased in the presence of PCM and nano PCM. Parveen et al. [131] measured the efficiency of heat sink by using nano-graphene and microencapsulated PCM. The particle size thickness was ranged from 5 to 20 nm and encapsulated PCM was paraffin-coated polyurethane. The results indicated that value of TC was increased from 0.192 to 0.379 W/mK by increasing NPs concentration from 0.5 to 3 wt%.

5.2.6 Energy Storage Material in Photovoltaic Thermal (PV/T) System

The PV panel converts very less amount of cosmic irradiance into electrical energy. The reasonable quantity of cosmic irradiance converts into heat due to which thermalization effects occur in panel and it gets heat up very much. As desires be the solar panel operates above than mentioned temperature [132]. Maintaining insulation as predictable, if the temperature of solar panel is extended, there is an increase in panel current, however a checked demotion in panel voltage [133]. The temperature of photovoltaic panels increases as the band gap breaks. As a result of this, the photo-generation rate of PV panel rises and a marginal augment in current [134]. However, the repeal saturation current also additions hastily with temperature and efficiency of PV panel decreases [135].

5.2.6.1 Photovoltaic System

The photovoltaic effect was first invented by French scientist Edmond Becquerel in 1839 and first silicon solar cell was made by group of researchers led by Chapin D. M at Bell Telephone Laboratories [136]. Photovoltaic cells are semiconductor devices that convert available energy into direct current electricity [137, 138]. This conversion process of available energy into electricity in PV cells is achieved by semiconductor materials from which cells are manufactured. The photovoltaic cell absorbs incident photons and converts them into electron–hole pair. Electrons jump from valence band to conduction band as energy absorbed by cell is equal or greater than band gap energy. Thus electron–hole pair generated that diffuses and separates the p–n junction of semiconductor due to electric field. The electrons move toward negative side as well as holes move toward positive side and finally electrons flow in external circuit and current is generated as illustrated in Fig. 5.13 [139].

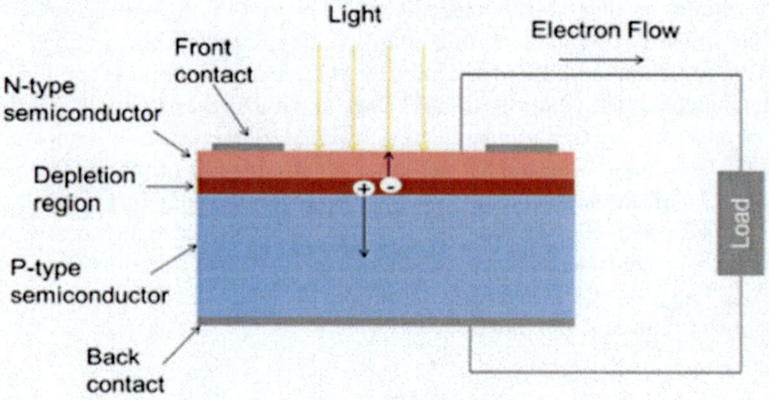

Fig. 5.13 P–n junction structure of photovoltaic cell [139]

The photovoltaic technology can be categorized into three main types as first generation, second generation and third generation. The first generation is fully commercialized generation contained crystalline silicon technology, which includes monocrystalline and polycrystalline silicon cell. The second generation contains photovoltaic thin film that includes amorphous silicon, cadmium telluride, indium copper selenide, indium and gallium selenide. The third generation is under consideration phase and includes organic photovoltaic, dye synthesized solar cell and gallium arsenide, indium phosphide, gallium antimonide [140] (Fig. 5.14).

5.2.6.2 Difficulties Associated with PV Panel Efficiency

Solar cells (made up of different materials: silicon material mostly used) are being used to alter the solar energy into electrical energy by photovoltaic process.

The PV modules are constructed by interconnecting the solar panels in parallel or series for getting the desired amount of power. Performance of solar module is not same at outdoor conditions as compared to controlled laboratory. Manufacturers provide the instructions of PV module at STC (standard test condition) or standard reporting conditions (that 1000 W/m^2, AM1.5 and 25 °C). These conditions are not found in the external conditions [141]. The performance of the PV module in actual or real conditions is the crucial issue for designer and installer for installing the PV system in definite location because the performance of PV module in outdoor is different at different climatic conditions. The difference is due to material of solar cell, meteorological parameter, different electrical and thermal characteristics, photovoltaic degradation and thermal annealing.

In tropical climates (such as Singapore), solar irradiance fluctuation was found very fast due to cloud coverage. Experimental results showed that among the five PV module of different technologies (monocrystalline, heterojunction Si wafer, a single junction, a double junction and micro morph Si), the efficiency of the wafer-based

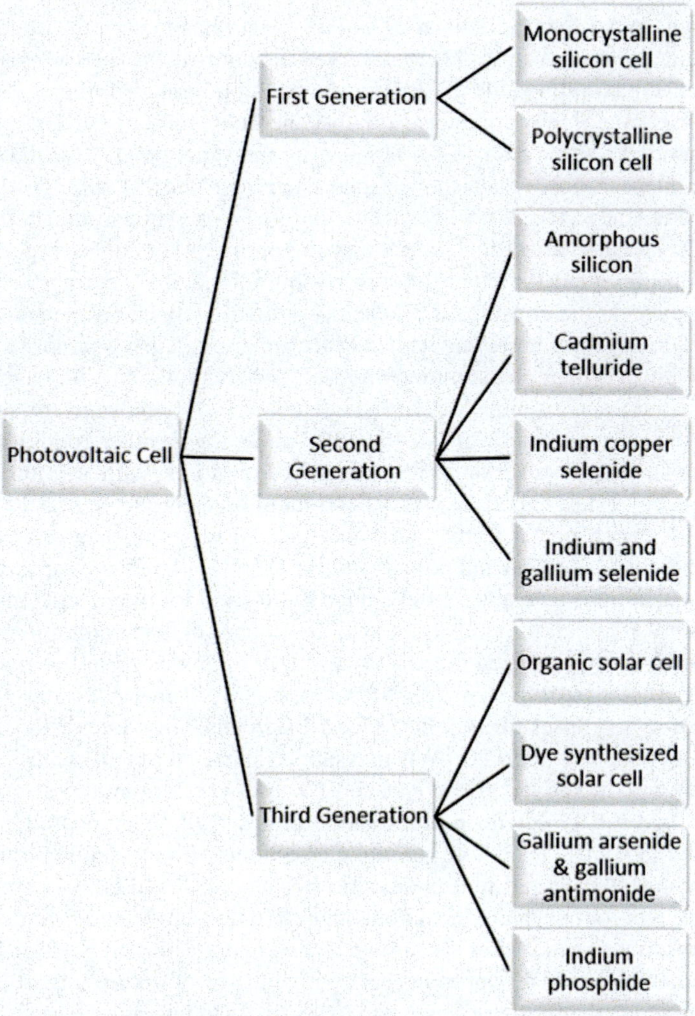

Fig. 5.14 Types of photovoltaic cell based on technologies [140]

silicon PV module was not much affected by the rapid fluctuation of solar irradiance while the mono crystalline and micro morph Si modules have shown higher ability at higher irradiance but dependent on the measure temperature [142]. Similarly, another study in tropical environment showed the degradation effect of mono and polycrystalline solar module in Dakar Senegal after a few years of operation. The degraded performance characteristics of crystalline solar module were Imax, Isc, Fill factor and P_{max} (losses were 0.22–2.96%/year). Congedo et al. [132] proclaimed the strong seasonal alteration in performance of 960 KW PV system (mainly consists of mono crystalline solar module) from March to October 2012 in the south-eastern

Italy. They found that the efficiency of PV system has maximum value of 17% in spring and minimum value of 15% observed in summer; PR (performance rate) had reached the maximum value of 86% in March and minimum value of 79% in June.

PANTIC et al. [143] conducted the feasible field study of performance of three equal monocrystalline solar modules (having the same power) for the whole year at different positions, result showed that each year vertical solar module has the highest efficiency (10.2%), performance ratio has also shown the similar result of 0.93, 0.91 and 0.86 at vertical, horizontal and optimum tilt angle, respectively. This study also presented the effect of surrounding temperature during the winter and summer days on the constituent efficiency, optimal inclined solar module has 30% lower efficiency in summer day while vertical solar module has opposite behavior in selected day of winter and summer whoever, horizontal module has stable behavior both in winter and summer days. Basoglu et al. [144] analyzed the performance of on grid photovoltaic power system (PPS) under the weather conditions of Izmit, Kocaeli of Northwest Turkey. PPS consists of mono crystalline, multi-crystalline and cadmium-telluride (Cd-Te) arrays and monitored for October 2013–December 2014. This experimental study shows that mean array efficiency (MAE) of Cd-Te was higher than the mono and polycrystalline arrays with highest capacity factor for all selected months and smaller deviations in PR ratio for mono and polycrystalline were observed higher as compared to Cd-Te in Izmit climatic conditions.

Eke et al. [145] analyzed the performance of 26 mono crystalline solar modules of grid-connected PV system (2.73 kW rated power), measured the performance from sunrise to sunset for whole the year by monitoring system. Results showed that PV system has 3856.90 kWh measured annual electricity with 7.3% annual efficiency and, moreover, grid-connected PV system has monthly electricity values varies between $\pm 12\%$ with performance ratio of 0.72. A comparative study was conducted by Canete et al. [146] of modules of flimsy film technologies that include the cadmium telluride and amorphous silicon in climatic conditions of southern Spain under sunlight for one year. The author suggested that Cd-Te and P-Si modules have better performance in winter season while a-Si and a-Si/μc-Si perform best during the summer season. The optimum best performance of amorphous solar modules was observed in the southern Spain due to low daily temperature loses. Abdel Kader et al. [147] analytically investigated that mono crystalline solar module has higher efficacy as compared to polycrystalline solar module in semi-arid climatic conditions of Jorden. Amin et al. [148] also investigated the production of mono crystalline, amorphous silicon, polycrystalline and copper indium in real conditions of Malaysia for consecutive 3 days and found that mono crystalline is not the best solution for Malaysian environment. Midgrated et al. [149] investigated the three different technologies under climatic of southern Norway and showed that the performance of mono crystalline solar module performed the best in terms of efficiency and electrical energy production. Predicting the output power is the leading issue for designer and installer in real conditions.

PV module performed best under low ambient temperature, but outdoor ambient temperature is not the same throughout the year due to which the efficiency of module is always less than the standard conditions [150]. The effect of PV module temperature mainly explained by many researchers for various PV modules and developed the various useful correlations of power/efficiency [151]. For reducing the PV module temperature, the cooling mechanism is required for achieving the high efficiency of photovoltaic solar module. Suwapaet et al. [152] confirmed and analyzed that output power of mono crystalline and amorphous photovoltaic modules is decreased to 20–35 and 3–20% as compared to standard conditions, respectively, under high operating temperature. The IV curve is the key parameter for collating the performance of PV module; therefore, it is most significant to measure the characteristics IV curve with care and high accuracy. Singh et al. [153] investigated the effect of module temperature on curve factor and open circuit voltage of silicon solar cell in temperature range 295–320 K. It was shown that shunt resistance was decreasing with increasing module temperature, furthermore, the series resistance depended on voltage also have an effect on the curve factor (CF).

The environmental elements such as dust accumulation, wind speed and direction, relative humidity etc. are also pretended the performance of photovoltaic solar module. Mekhlief et al. [154] studied the compound effect of wind, dust and humidity on the PV module operation. It was shown that efficiency of solar module decreases with an increase in the dust deposition on solar module, but efficiency is increasing with increasing wind speed, increasing wind causes to lowering humidity that leads to increase the efficiency. Jiang et al. [155] investigated the effect of dust deposition using test chamber and found a decline in module efficiency up to 26% for dust accumulation of 22 g/m^2. Pili et al. [156] conducted the 8-month experimental study in KSA and reported that the efficacy of solar module was reduced up to 32%.

Apart from that, PV designer and installer have required the simple, accurate, reasonable accuracy and easy to apply methods for predicting the output power and energy yield of PV module. Fuentes et al. [157] evaluate the performance of solar module with mono and polycrystalline structure in Mediterranean climate for all the year. The obtained result showed that predicting the output power of mono and polycrystalline solar modules by using the AMPP (approximate maximum power point), Ostarwald and constant fill factor methods are best suited in Mediterranean climates. Under the Climate of Madrid and Jaen (Spain) [158], Osterwald and constant fill factor method have performed best in terms of power with RMSE of 3.8 and 4.5% for Cadmium telluride (Cd-Te) and Copper indium gallium selenite sulfide (CIGS) respectively [159].

Some previous studies showed that Pakistan has the immense potential for the implementation of solar photovoltaic system due to having high insolation of 15–21 MJ/M^2 with annually 3000–3300 sun hours and average daily long sunshine hours (7–8 h) [160]. The average highest solar energy received in the month of June and July as reported by Ulfat et al. [161]. Therefore, Pakistan is the favorable country for the installing of PV system in most areas.

Like as Pakistan, there is broad literature accessible on the electricity demand in every country of world. With increasing demand of electricity, the cost of energy increases between 30 and 40% [162]. Due to a weak correlation between electricity price and its demand, a strategy built on price escalation may not work toward curtailing demand. To increase the demand of electricity with the passage of time, the integrated and efficient systems should be used and to shift the load from conventional energy resources to renewable energy resources. For this purpose, solar energy is inexhaustible, most abundant and clean of all the renewable energy resources till date. In the present rate of all the energy exhaustion, solar energy was found to be many times superior, which is captured by the earth is about 1.8×10^{11} MW [163].

It is well known that the increase of 1 °C of panel temperature is reduced 0.5% panel ability. So, it is needed to work properly at low temperatures to maintain electrical efficiency of PV cell at reasonable level. For that reason, scientists and engineers are watching for very low-cost cooling system for the solar panels [164]. At Energy Studies Institute, NUS (National University of Singapore) experiments are done on Four 55-W polycrystalline solar cells with using and without using of active cooling system to produce electricity. The result shows that solar cells can only achieve efficacy of 8–9% without active cooling where the temperature of the cell was as elevated as 68 °C. But, the temperature decreased noticeably to 38 °C, which defined efficiency of solar cells prominent increase from 12 and 14% with using of active cooling conditions, where airflow rate was held 0.05 kg/s.

A novel solar panel assimilated with water tubes embed in functionally graded material and photovoltaic cells are fictional. The FGM layer can avert heat loss to the substrate by heat transfer from the PV cells to the water tubes. The hybrid solar panel is working at fairly low temperatures, which shows eager performance with PV cells [165]. It is acknowledged that up to 80% of the solar irradiation is absorbed by PV cells. 5–20% of the total incident energy on PV cells is converted into electrical energy. Due to this reason, a large amount of incident energy causes excessive heating, as in sunny days, the temperature of PV cells can go as high as 37 °C above air temperature [166, 167].

PV cell converts feeble fragment of solar irradiance into electrical energy and considerable quantity of radiations converts into heat, due to which cell warming occurs. Hence, the cell starts operating above surrounding temperature. Keeping solar radiations level as constant, if the temperature of the panel raised, then the minimal increase in the PV cell current occurs but a marked alleviation in cell voltage also occurs. When band gap breaks sharply, the PV panel's temperature rises. So, there is an increase in photo-generation rate and thus, PV panel current rises but, at the same time, the repeal saturation current rises briskly due to this increase in panel temperature due to which the conversion efficacy of PV panel from solar to electrical energy reduces [168]. The decrement of cell efficiency with an increase in temperature is shown in Fig. 5.15.

Mainly the measured power of the PV module depends on many elements like as temperature, how much solar irradiance, location or angle of PV module, and some other environmental parameters like humidity, dust, wind velocity, direction of wind, etc. Therefore, before the installation of PV module in specific climate, installer and designer have required the accurate.

PV panel is one of the key solar energy systems. However, its performance strongly depends upon surface temperature. Best efficiency of PV panel is in the range of 15–18% and it decreases as the temperature of PV panel increases from a specific

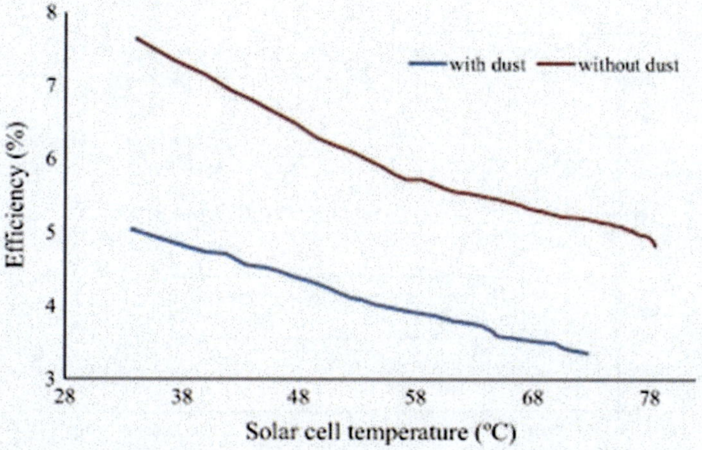

Fig. 5.15 Decrement of temperature with increasing temperature of PV panel [169]

limit, i.e. 25 °C [170]. The performance of PV panel can be further increased by thermal management through different cooling techniques such as passive cooling using phase change material.

5.2.6.3 Addition of PCM/nano PCM in PV/T Systems

The addition of PCM in PV panel enhances the efficiency of system by absorbing waste heat and thus improves the performance of panels. The PCMs have high latent heat, which resulted in absorbing more amount of heat compared to air and water and it is more efficient in terms of absorbing heat. The usage of PCM in cooling of PV module was first examined by Stultz and Wen [171] in 1977. The results showed that PCM was a useful technique for cooling of PV module and efficiency was enhanced by absorbing extra heat. Figure 5.16 shows the PV panel evaluation with and without PCM [172].

Smith et al.[173] used PCM as heat sink to improve the performance of PV panel by reducing the temperature of system. Various PCMs having melting temperature 0–50 °C were investigated to measure its impact on system. The results indicated that by using PCM, the cooling of system was improved and annual PV output enhancement is illustrated in Fig. 5.17 and at some locations, the enhancement was 6%.

Stropnik and Stritih [174] improved the performance of PV/T system by adding PCM and enhanced the efficiency of system. The simulation on TRYSYS as well as experimentation (shown in Fig. 5.18) was done to measure heat extraction from PV system. The results indicated that temperature difference of PV surface panel with PCM and without PCM was 35.6 °C and maximum electrical efficiency was 13%.

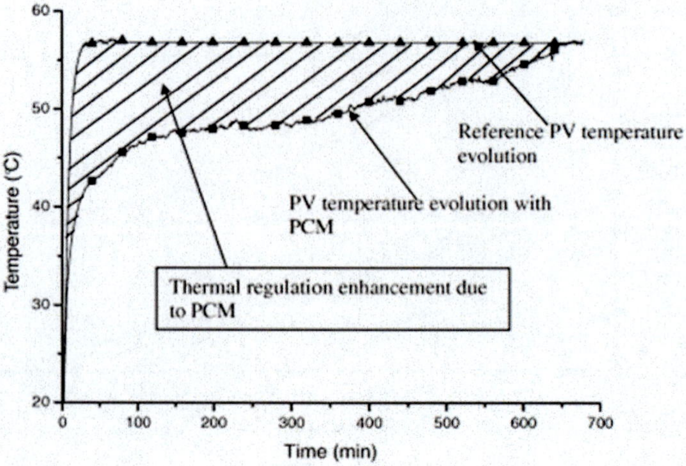

Fig. 5.16 PV panel evaluation with and without PCM [172]

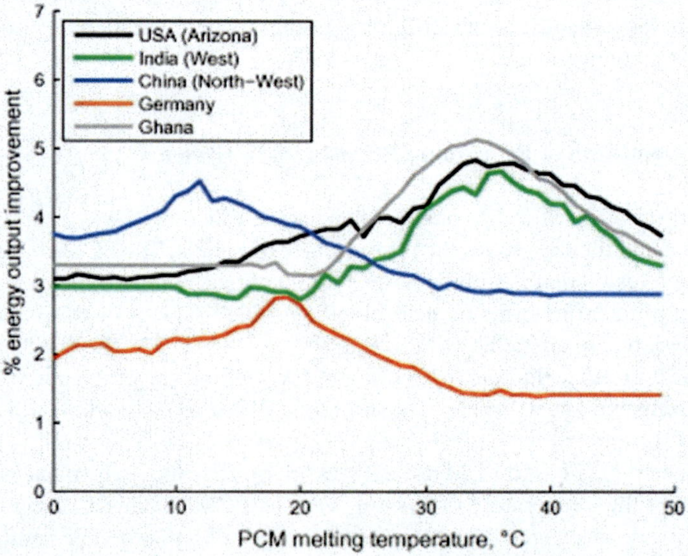

Fig. 5.17 The increment in PV output as a function melting temperature of PCM [173]

Maiti et al. [175] investigated PV-PCM system having paraffin wax used as PCM. The paraffin wax has melting temperature 56–58 °C for indoor and outdoor experimentation. The experiments were carried out and results were concluded that temperature reduced from 90 to 65–68 °C for indoor conditions while 72–62 °C for outdoor conditions. The power output increment was 55% using PV-PCM system. Biwole et al. [176] examined CFD analysis for measuring heat transfer using PCM inserted

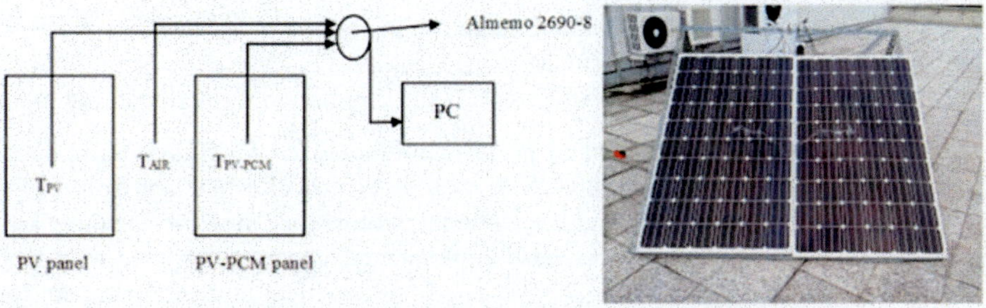

Fig. 5.18 Schematic diagram of experimental setup (left) and front side of PV and PV-PCM panel measurement (right) [174]

at back side of panel. The fins having 4-mm width were used for thermal analysis and system efficiency was studied along fins. The results indicated that by adding PCM, the temperature of panel was maintained to 40 °C for 80 min.

Haung et al. [177] investigated PCM-based PV/T system having metal fin configuration for cooling of PV system. The three PCMs, namely, RT27, RT35 and Waksol A were used by 85% to improve efficiency. It was observed that by using RT27 and fins, there was significant reduction in increment of temperature. Browne et al. [178] performed study on PCM-PV/T, PV/T-water and non-cooled PV panel. The stainless steel contained was used with tubing to pass water and PCM. The results concluded that PCM-PV system exhibited high heat storage compared to PV/T-water under same conditions. Browne et al. [179] also studied potential of PCM for efficient performance of PV panel, concentrated PV and building integrated PV panel. The authors work revealed that PCM-based system was a better technique for cooling system. Aktin et al. [180] studied four different techniques for thermal regulations of PV panels. The two monocrystalline silicon solar panels were under consideration. The four different configurations were investigated and the increment in overall efficiency of PV panel was 12.9% with last case having PCM and finned.

The efficiency of PV panels is still low by integrating with only PCM, which can be enhanced by various methods. One of the major techniques is to add NPs in PCM, which results in increment in thermal conductivity of PCM. The addition of high value of "k" of NPs enhances the performance capacity of base PCM. Al-Waeli et al. [181] investigated nano PCM and nanofluid-based photovoltaic thermal system to improve the overall efficiency of system. The SiC NPs were used in paraffin PCM and nanofluid (Water + SiC) at flow rate of 0.17 kg/s was used for improvement in value of "k" and to enhance thermal efficiency. The experiments were conducted under the outdoor conditions of Selangor, Malaysia. The results indicated that PV cell temperature was reduced by 30 °C, open circuit voltage enhanced from 11–13 to 20–21 V. The power was increased from 61.1 to 120.7 W, electrical efficiency improved from 7.1 to 13.7% and thermal efficiency reached up to 72%.

Al-Waeli et al. [182] also compared three different types of PV/T system with conventional PV system. The three various cooling techniques were proposed:

- Tank filled with water and water circulation in cooling pipes
- Tank filled with nano PCM (PCM + nano SiC)
- Tank filled with nanofluid (water + SiC) and circulation through cooling pipes.

The results indicated that nano PCM nanofluid-based PV/T system enhanced electric current from 3.69 to 4.04 A as well as electrical efficiency from 8.07 to 13.32% when compared with the conventional photovoltaic system. The three artificial neutral network methods were also implemented, and results agreed with experimental results. Al-Waeli et al. [183] carried out technical and economic study of PV/T system containing tank having nano PCM attached with PV panel, nanofluids inside the pipe, which passed through nano PCM tank at 0.175 kg/s as represented in Fig. 5.19. The results concluded that the efficiency of system was increased from 7.1 to 13.7% as compared to conventional system. The output water temperature was 39.52 °C and system was economically viable.

Dhaidan et al. [184] carried out tests experimentally as well as numerically to check the melting of nano PCM under constant heat flux. The value of "k" and melting rate was enhanced by adding NPs in base PCM. The rate of melting was higher at low value of NPs loading, which resulted in higher energy storage and saved cost in comparison with higher NPs loading. Groulx [185] investigated melting and freeing of nano PCM in a container having metallic fins in it. It was concluded that base pcm with metallic fins and NPs was performed much better in terms of thermal storage and HT rate.

Parameshwarah et al. [186] conducted the comparison of nano PCM with pure PCM in building applications for thermal storage. The nano PCM contained silver

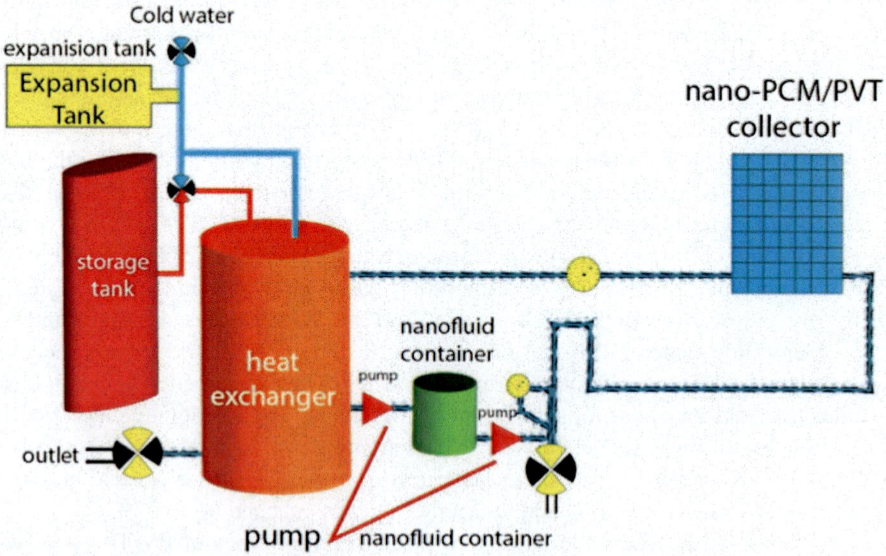

Fig. 5.19 Experimental setup proposed by Al-Waeli et al. [183]

and titanium NPs in base PCM and it was observed that freezing time of nano PCM was lower than pure PCM, which resulted in increment in value of “k” and thermal storage capacity. Amin et al. [187] enhanced the value of “k” of base PCM beeswax by adding graphene NPs, which was later used in building applications. It was observed that the value of “k” increased to 2.8 W/m.K and latent heat enhanced by 22.5% having 0.3 wt% of beeswax/graphene. Luo et al. [188] prepared form-stable composite PCM having mixture of capric acid, palmitic acid, steric acid and NPs of silica. It was observed that nano PCM having phase change temperature of 17.16–26 °C was suitable for maintaining indoor temperature of buildings. Sharma et al. [189] investigated the cooling effect of building integrated concentrated photovoltaic (BICPV) by using micro fins/PCM and nano PCM. The thermal efficiency was first with un-fins and with fins and then checked with PCM and nano PCM. The reduction in temperature in the center of system was 10.7 °C by using micro-fins with PCM and 12.5 °C by using micro-fins with nano PCM. The decrement in temperature without micro-fins of PCM and nano PCM was 9.6 and 11.2 °C respectively, compared to natural convection case. Ma et al. [190] presented nano PCM-based thermal storage in ceiling ventilation system integrated with PV/T system. The RT24 PCM was mixed with Cu NPs and conclusion was drawn. The results depicted that nano PCM has higher melting and solidification time was improved compared to pure PCM. The 8.3% of more heat was absorbed and 25.1% of more heat was released compared to pure PCM. Lin and Al-Kayiem [191] investigated the 20-nm five samples of copper nanoparticle dispersed into paraffin wax for solar energy thermal storage. The result demonstrated that for all dispersed weight of Cu nanoparticles, the thermal conductivity of Cu-PCM nanocomposite was increased (Table 5.4).

The chapter concludes the applications of advanced energy storage materials in a wide range of areas of thermal applications that include buildings, solar energy, waste heat recovery, seawater desalination, electronic cooling and photovoltaic thermal systems. These advanced materials have broadly used in these areas, which resulted in performance enhancement greatly. The researchers should continuously explore the properties of these materials to broaden their area of applications. There should be more areas of applications exploit in the field of thermal storage system for these materials for future research. More work is to be needed to explore the new types of energy storage materials for applications in various fields. The properties related to advanced storage materials should be considered further for utilization in more applications with much better execution.

Table 5.4 Summary of various authors on PV/T systems using energy storage materials

References	PCM	NPs	Nano concentration	Results
[192]	82 wt% co-conut oil and 18 wt% sunflower oil	Boehmite (AlO(OH))	0.009 w/w	Efficiency increased from 29 to 48% by nano-PCM at 690 W/m^2. Along 25 °C temperature was reduced when compared to without PCM
[193]	Paraffin (RT35HC)	Al_2O_3	0.11, 0.22, 0.33, 0.44, 0.55, 0.66, 0.77 (vol%)	The temperature was reduced from 1 to 4.5 °C from 0.11 to 0.77 vol% of NPs at 820 W/m^2
[194]	Coconut oil	CuO	0, 2, 4, 5 (vol%)	The melting temperature of nano PCM was improved
[195]	Paraffin	Cu	2 wt%	The 14.2% increment in thermal conductivity by adding NPs in base PCM
[196]	Paraffin	Alumina, Titania, silica	1, 2, 3 (wt%)	Titania showed more effective in terms of energy storage
[197]	Erythritol	Graphite particles	15 vol%	Thermal conductivity increased by 640% compared to pure PCM by adding graphite particles
[198]	Barium chloride	Titania particles	0.07, 0.13, 0.25, 0.50 (wt%)	The 12.76% enhancement in thermal conductivity at 0.50 wt% of Titania NPs was observed
[199]	Dedecanol	MWCNTs	1 and 2 (wt%)	The 4.6 and 11.0% enhancement in thermal conductivity by addition of 1 and 2 wt% of CNTs, respectively
[200]	Paraffin wax	CNTs	1, 2, 5, 10 (wt%)	The 5.9–24.4% increment in thermal conductivity by using CNTs in base PCM

(continued)

Table 5.4 (continued)

References	PCM	NPs	Nano concentration	Results
[201]	Paraffin wax	graphite	1, 4, 7, 10 (wt%)	The 4.3–167.5% enhancement in thermal conductivity of base PCM by adding graphite NPs
[202]	Paraffin wax	Al_2O_3, ZnO_2, SiC	0.1, 0.5, 1, 2, 3, 4 (wt%)	Thermal conductivity improved by 3.3, 1.8 and 4.2% with the addition of 1% nano-Al_2O_3, nano-ZnO_2 and nano-SiC, respectively
[203]	Paraffin wax	MWCNTs	0.2, 0.5% by mass	The maximum increment in value of "k" was observed 0.203 W/mK at 0.5% mass of MWCNTs

References

1. M. Sokolov, Y. Keizman, Performance indicators for solar pipes with phase change storage. Sol. Energy **47**, 339–346 (1991)
2. H.E.S. Fath, Technical assessment of solar thermal energy storage technologies. Renew. Energy **14**, 35–40 (1998)
3. E.-B.S. Mettawee, G.M.R. Assassa, Experimental study of a compact PCM solar collector. Energy **31**, 2958–2968 (2006)
4. A. Shukla, D. Buddhi, R.L. Sawhney, Solar water heaters with phase change material thermal energy storage medium: A review. Renew. Sustain. Energy Rev. **13**, 2119–2125 (2009)
5. M. Iten, S. Liu, A. Shukla, A review on the air-PCM-TES application for free cooling and heating in the buildings. Renew. Sustain. Energy Rev. **61**, 175–186 (2016)
6. E. Osterman, V. Butala, U. Stritih, PCM thermal storage system for 'free'heating and cooling of buildings. Energy Build. **106**, 125–133 (2015)
7. F. Guarino, A. Athienitis, M. Cellura, D. Bastien, PCM thermal storage design in buildings: experimental studies and applications to solaria in cold climates. Appl. Energy **185**, 95–106 (2017)
8. B. Nie, X. She, Z. Du, C. Xie, Y. Li, Z. He, Y. Ding, System performance and economic assessment of a thermal energy storage based air-conditioning unit for transport applications. Appl. Energy **251**, 113254 (2019)
9. S. Gharbi, S. Harmand, S. Ben Jabrallah, Parametric study on thermal performance of PCM heat sink used for electronic cooling, in *Exergy a Better Environment and Improved Sustainability*, vol. 1 (Springer, 2018), pp. 243–256
10. R. Kothari, P. Mahalkar, S.K. Sahu, S.I. Kundalwal, Experimental investigations on thermal performance of PCM based heat sink for passive cooling of electronic components, in *ASME 2018 16th International Conference on Nanochannels, Microchannels, Minichannels, American Society of Mechanical Engineers*, pp. V001T11A005–V001T11A005 (2018)
11. H. Usman, H.M. Ali, A. Arshad, M.J. Ashraf, S. Khushnood, M.M. Janjua, S.N. Kazi, An experimental study of PCM based finned and un-finned heat sinks for passive cooling of electronics. Heat Mass Transf. **54**, 3587–3598 (2018)

12. S.S. Chougule, V.V. Nirgude, S.P. Shewale, A.T. Pise, S.K. Sahu, H. Shah, Application of paraffin based nanocomposite in heat pipe module for electronic equipment cooling. Mater. Today Proc. **5**, 23333–23338 (2018)
13. A. Moldgy, R. Parameshwaran, Study on thermal energy storage properties of organic phase change material for waste heat recovery applications. Mater. Today Proc. **5**, 16840–16848 (2018)
14. K. Merlin, J. Soto, D. Delaunay, L. Traonvouez, Industrial waste heat recovery using an enhanced conductivity latent heat thermal energy storage. Appl. Energy **183**, 491–503 (2016)
15. A.I. Fernández, C. Barreneche, L. Miró, S. Brückner, L.F. Cabeza, Thermal energy storage (TES) systems using heat from waste, in *Advanced Thermal Energy Storage Technology* (Elsevier, 2015), pp. 479–492
16. M. Jaworski, R. Wnuk, M. Cieślak, B. Goetzendorf-Grabowska, Experimental investigation and mathematical modelling of thermal performance characteristics of textiles incorporating phase change materials (PCMs), in: *Environmental Engineering. Proceedings of the International Conference on Environmental Engineering* (ICEE, Vilnius Gediminas Technical University, Department of Construction Economics, 2017), pp. 1–9
17. X. Liu, Y. Lou, Preparation of microencapsulated phase change materials by the sol-gel process and its application on textiles. Fibres Text. East. Eur. (2015)
18. P. Zhang, J. Li, L. Lv, Y. Zhao, L. Qu, Vertically aligned graphene sheets membrane for highly efficient solar thermal generation of clean water. ACS Nano **11**, 5087–5093 (2017)
19. P. Zhang, Q. Liao, T. Zhang, H. Cheng, Y. Huang, C. Yang, C. Li, L. Jiang, L. Qu, High throughput of clean water excluding ions, organic media, and bacteria from defect-abundant graphene aerogel under sunlight. Nano Energy **46**, 415–422 (2018)
20. L. Zhou, Y. Tan, J. Wang, W. Xu, Y. Yuan, W. Cai, S. Zhu, J. Zhu, 3D self-assembly of aluminium nanoparticles for plasmon-enhanced solar desalination. Nat. Photonics **10**, 393 (2016)
21. X. Li, W. Xu, M. Tang, L. Zhou, B. Zhu, S. Zhu, J. Zhu, Graphene oxide-based efficient and scalable solar desalination under one sun with a confined 2D water path. Proc. Natl. Acad. Sci. **113**, 13953–13958 (2016)
22. K.-K. Liu, Q. Jiang, S. Tadepalli, R. Raliya, P. Biswas, R.R. Naik, S. Singamaneni, Wood–graphene oxide composite for highly efficient solar steam generation and desalination. ACS Appl. Mater. Interfaces **9**, 7675–7681 (2017)
23. H. Ren, M. Tang, B. Guan, K. Wang, J. Yang, F. Wang, M. Wang, J. Shan, Z. Chen, D. Wei, Hierarchical graphene foam for efficient omnidirectional solar–thermal energy conversion. Adv. Mater. **29**, 1702590 (2017)
24. J. Wang, Y. Li, L. Deng, N. Wei, Y. Weng, S. Dong, D. Qi, J. Qiu, X. Chen, T. Wu, High-performance photothermal conversion of narrow-bandgap Ti_2O_3 nanoparticles. Adv. Mater. **29**, 1603730 (2017)
25. Z. Yin, H. Wang, M. Jian, Y. Li, K. Xia, M. Zhang, C. Wang, Q. Wang, M. Ma, Q. Zheng, Extremely black vertically aligned carbon nanotube arrays for solar steam generation. ACS Appl. Mater. Interfaces **9**, 28596–28603 (2017)
26. T.R. Shah, H.M. Ali, Applications of hybrid nanofluids in solar energy, practical limitations and challenges: a critical review. Sol. Energy **183**, 173–203 (2019)
27. C. Chen, Y. Li, J. Song, Z. Yang, Y. Kuang, E. Hitz, C. Jia, A. Gong, F. Jiang, J.Y. Zhu, Highly flexible and efficient solar steam generation device. Adv. Mater. **29**, 1701756 (2017)
28. X. Hu, W. Xu, L. Zhou, Y. Tan, Y. Wang, S. Zhu, J. Zhu, Tailoring graphene oxide-based aerogels for efficient solar steam generation under one sun. Adv. Mater. **29**, 1604031 (2017)
29. L. Cui, P. Zhang, Y. Xiao, Y. Liang, H. Liang, Z. Cheng, L. Qu, High rate production of clean water based on the combined photo-electro-thermal effect of graphene architecture. Adv. Mater. **30**, 1706805 (2018)
30. C. Chen, L. Zhou, J. Yu, Y. Wang, S. Nie, S. Zhu, J. Zhu, Dual functional asymmetric plasmonic structures for solar water purification and pollution detection. Nano Energy **51**, 451–456 (2018)

31. V. Dao, H. Choi, Carbon-based sunlight absorbers in solar-driven steam generation devices. Glob. Chall. **2**, 1700094 (2018)
32. F. Zhao, X. Zhou, Y. Shi, X. Qian, M. Alexander, X. Zhao, S. Mendez, R. Yang, L. Qu, G. Yu, Highly efficient solar vapour generation via hierarchically nanostructured gels. Nat. Nanotechnol. **13**, 489 (2018)
33. X. Yang, Y. Yang, L. Fu, M. Zou, Z. Li, A. Cao, Q. Yuan, An ultrathin flexible 2D membrane based on single-walled nanotube–MoS2 hybrid film for high-performance solar steam generation. Adv. Funct. Mater. **28**, 1704505 (2018)
34. H.M. Ali, T.R. Shah, H. Babar, Z.A. Khan, Application of nanofluids for thermal management of photovoltaic modules: a review, in: *Microfluid. Nanofluidics, IntechOpen* (2018)
35. A. Abhat, Low temperature latent heat thermal energy storage: heat storage materials. Sol. Energy **30**, 313–332 (1983)
36. I. Dincer, S. Dost, X. Li, Performance analyses of sensible heat storage systems for thermal applications. Int. J. Energy Res. **21**, 1157–1171 (1997)
37. V.V. Tyagi, S.C. Kaushik, S.K. Tyagi, T. Akiyama, Development of phase change materials based microencapsulated technology for buildings: a review. Renew. Sustain. Energy Rev. **15**, 1373–1391 (2011)
38. M. Ravikumar, P.S.S. Srinivasan, Phase change material as a thermal energy storage material for cooling of building. J. Theor. Appl. Inf. Technol. 4 (2008)
39. S.E. Kalnæs, B.P. Jelle, Phase change materials and products for building applications: a state-of-the-art review and future research opportunities. Energy Build. **94**, 150–176 (2015)
40. T. Khadiran, M.Z. Hussein, Z. Zainal, R. Rusli, Advanced energy storage materials for building applications and their thermal performance characterization: a review. Renew. Sustain. Energy Rev. **57**, 916–928 (2016)
41. I. Marco, Seminar on phase change materials and innovation products—Brianza Plastica, Beijing, China, Oct 20 2005
42. K. Peippo, P. Kauranen, P.D. Lund, A multicomponent PCM wall optimized for passive solar heating. Energy Build. **17**, 259–270 (1991)
43. D.A. Neeper, Thermal dynamics of wallboard with latent heat storage. Sol. Energy. **68**, 393–403 (2000)
44. D. Heim, J.A. Clarke, Numerical modelling and thermal simulation of PCM–gypsum composites with ESP-r. Energy Build. **36**, 795–805 (2004)
45. D. Rozanna, A. Salmiah, T.G. Chuah, R. Medyan, S.Y. Thomas Choong, M. Sa ari, A study on thermal characteristics of phase change material (PCM) in gypsum board for building application. J. Oil Palm Res. **17**, 41 (2005)
46. H. Kaasinen, The absorption of phase change substances into commonly used building materials. Sol. Energy Mater. Sol. Cells. **27**, 173–179 (1992)
47. D. Feldman, D. Banu, D. Hawes, E. Ghanbari, Obtaining an energy storing building material by direct incorporation of an organic phase change material in gypsum wallboard. Sol. Energy Mater. **22**, 231–242 (1991)
48. Y. Zhang, G. Zhou, K. Lin, Q. Zhang, H. Di, Application of latent heat thermal energy storage in buildings: state-of-the-art and outlook. Build. Environ. **42**, 2197–2209 (2007)
49. M. Zhang, M.A. Medina, J.B. King, Development of a thermally enhanced frame wall with phase-change materials for on-peak air conditioning demand reduction and energy savings in residential buildings. Int. J. Energy Res. **29**, 795–809 (2005)
50. P. Schossig, H.-M. Henning, S. Gschwander, T. Haussmann, Micro-encapsulated phase-change materials integrated into construction materials. Sol. Energy Mater. Sol. Cells. **89**, 297–306 (2005)
51. A. Oliver, Thermal characterization of gypsum boards with PCM included: thermal energy storage in buildings through latent heat. Energy Build. **48**, 1–7 (2012)
52. E. Rodriguez-Ubinas, L. Ruiz-Valero, S. Vega, J. Neila, Applications of phase change material in highly energy-efficient houses. Energy Build. **50**, 49–62 (2012)
53. J.-F. Su, X.-Y. Wang, S.-B. Wang, Y.-H. Zhao, Z. Huang, Fabrication and properties of microencapsulated-paraffin/gypsum-matrix building materials for thermal energy storage. Energy Convers. Manag. **55**, 101–107 (2012)

54. T. Toppi, L. Mazzarella, Gypsum based composite materials with micro-encapsulated PCM: experimental correlations for thermal properties estimation on the basis of the composition. Energy Build. **57**, 227–236 (2013)
55. C. Castellon, M. Nogués, J. Roca, M. Medrano, L.F. Cabeza, *Microencapsulated Phase Change Materials (PCM) for Building Applications* (ECOSTOCK, NJ, 200)
56. N.A. Yahay, H. Ahmad, Numerical investigation of indoor air temperature with the application of PCM gypsum board as ceiling panels in buildings. Procedia Eng. **20**, 238–248 (2011)
57. S. Scalat, D. Banu, D. Hawes, J. Parish, F. Haghighata, D. Feldman, Full scale thermal testing of latent heat storage in wallboard. Sol. Energy Mater. Sol. Cells. **44**, 49–61 (1996)
58. F. Kuznik, J. Virgone, J.-J. Roux, Energetic efficiency of room wall containing PCM wallboard: a full-scale experimental investigation. Energy Build. **40**, 148–156 (2008)
59. L.F. Cabeza, C. Castellon, M. Nogues, M. Medrano, R. Leppers, O. Zubillaga, Use of microencapsulated PCM in concrete walls for energy savings. Energy Build. **39**, 113–119 (2007)
60. P. Arce, C. Castellón, A. Castell, L.F. Cabeza, Use of microencapsulated PCM in buildings and the effect of adding awnings. Energy Build. **44**, 88–93 (2012)
61. A.M. Thiele, G. Sant, L. Pilon, Diurnal thermal analysis of microencapsulated PCM-concrete composite walls. Energy Convers. Manag. **93**, 215–227 (2015)
62. H. Inaba, P. Tu, Evaluation of thermophysical characteristics on shape-stabilized paraffin as a solid-liquid phase change material. Heat Mass Transf. **32**, 307–312 (1997)
63. M. Xiao, B. Feng, K. Gong, Preparation and performance of shape stabilized phase change thermal storage materials with high thermal conductivity. Energy Convers. Manag. **43**, 103–108 (2002)
64. Y. Hong, G. Xin-Shi, Preparation of polyethylene–paraffin compound as a form-stable solid-liquid phase change material. Sol. Energy Mater. Sol. Cells. **64**, 37–44 (2000)
65. M. Xiao, B. Feng, K. Gong, Thermal performance of a high conductive shape-stabilized thermal storage material. Sol. Energy Mater. Sol. Cells. **69**, 293–296 (2001)
66. M. Li, Z. Wu, M. Chen, Preparation and properties of gypsum-based heat storage and preservation material. Energy Build. **43**, 2314–2319 (2011)
67. K. Biswas, J. Lu, P. Soroushian, S. Shrestha, Combined experimental and numerical evaluation of a prototype nano-PCM enhanced wallboard. Appl. Energy **131**, 517–529 (2014)
68. Y.P. Zhang, K.P. Lin, R. Yang, H.F. Di, Y. Jiang, Preparation, thermal performance and application of shape-stabilized PCM in energy efficient buildings. Energy Build. **38**, 1262–1269 (2006)
69. G. Zhou, Y. Zhang, K. Lin, W. Xiao, Thermal analysis of a direct-gain room with shape-stabilized PCM plates. Renew. Energy **33**, 1228–1236 (2008)
70. Y. Sun, S. Wang, F. Xiao, D. Gao, Peak load shifting control using different cold thermal energy storage facilities in commercial buildings: a review. Energy Convers. Manag. **71**, 101–114 (2013)
71. G. Kumaresan, R. Sridhar, R. Velraj, Performance studies of a solar parabolic trough collector with a thermal energy storage system. Energy **47**, 395–402 (2012)
72. C. Prieto, R. Osuna, A.I. Fernández, L.F. Cabeza, Thermal storage in a MW scale. Molten salt solar thermal pilot facility: plant description and commissioning experiences. Renew. Energy **99**, 852–866 (2016)
73. S. Guillot, A. Faik, A. Rakhmatullin, J. Lambert, E. Veron, P. Echegut, C. Bessada, N. Calvet, X. Py, Corrosion effects between molten salts and thermal storage material for concentrated solar power plants. Appl. Energy **94**, 174–181 (2012)
74. V. Zipf, A. Neuhäuser, C. Bachelier, R. Leithner, W. Platzer, Assessment of different PCM storage configurations in a 50 MWel CSP plant with screw heat exchangers in a combined sensible and latent storage–simulation results. Energy Procedia **69**, 1078–1088 (2015)
75. M.H. Mahfuz, A. Kamyar, O. Afshar, M. Sarraf, M.R. Anisur, M.A. Kibria, R. Saidur, I. Metselaar, Exergetic analysis of a solar thermal power system with PCM storage. Energy Convers. Manag. **78**, 486–492 (2014)

76. K. Bhagat, S.K. Saha, Numerical analysis of latent heat thermal energy storage using encapsulated phase change material for solar thermal power plant. Renew. Energy **95**, 323–336 (2016)
77. S.C. Lin, H.H. Al-Kayiem, Evaluation of copper nanoparticles–Paraffin wax compositions for solar thermal energy storage. Sol. Energy **132**, 267–278 (2016)
78. D.K. Singh, S. Suresh, H. Singh, B.A.J. Rose, S. Tassou, N. Anantharaman, Myo-inositol based nano-PCM for solar thermal energy storage. Appl. Therm. Eng. **110**, 564–572 (2017)
79. P.A. Galione, C.D. Pérez-Segarra, I. Rodríguez, S. Torras, J. Rigola, Multi-layered solid-PCM thermocline thermal storage for CSP. Numerical evaluation of its application in a 50 MWe plant. Sol. Energy **119**, 134–150 (2015)
80. I. IEA-ETSAP, Thermal energy storage: technology brief e17 (2013)
81. L. Miró, J. Gasia, L.F. Cabeza, Thermal energy storage (TES) for industrial waste heat (IWH) recovery: a review. Appl. Energy **179**, 284–301 (2016)
82. J. Selvaraj, M. Thenarasu, S. Aravind, P. Ashok, Waste heat recovery from castings and scrap preheating by recovered heat using an intermediate heat transfer medium, in: *Applied Mechanics and Materials* (Trans Tech Publications 2015), pp. 776–781
83. N. Maruoka, K. Sato, J. Yagi, T. Akiyama, Development of PCM for recovering high temperature waste heat and utilization for producing hydrogen by reforming reaction of methane. ISIJ Int. **42**, 215–219 (2002)
84. L. Zhang, T. Akiyama, How to recuperate industrial waste heat beyond time and space. Int. J. Exergy **6**, 214–227 (2009)
85. V. Pandiyarajan, M. Chinnappandian, V. Raghavan, R. Velraj, Second law analysis of a diesel engine waste heat recovery with a combined sensible and latent heat storage system. Energy Policy **39**, 6011–6020 (2011)
86. M. Chinnapandian, V. Pandiyarajan, A. Prabhu, R. Velraj, Experimental investigation of a cascaded latent heat storage system for diesel engine waste heat recovery, Energy Sources, Part A Recover. Util. Environ. Eff. **37**, 1308–1317 (2015)
87. T. Steinparzer, M. Haider, A. Fleischanderl, A. Hampel, G. Enickl, F. Zauner, Heat exchangers and thermal energy storage concepts for the off-gas heat of steelmaking devices. J. Phys. Conf. Ser. 12158 (2012). IOP Publishing
88. S.S. Prabu, M.A. Asokan, A study of waste heat recovery from diesel engine exhaust using phase change material. Int. J. Chem. Tech. Res. **8**, 711–717 (2015)
89. N. Gopal, R. Subbarao, V. Pandiyarajan, R. Velraj, Thermodynamic analysis of a diesel engine integrated with a PCM based energy storage system. Int. J. Thermodyn. **13**, 15–21 (2010)
90. P. Kauranen, T. Elonen, L. Wikström, J. Heikkinen, J. Laurikko, Temperature optimisation of a diesel engine using exhaust gas heat recovery and thermal energy storage (diesel engine with thermal energy storage). Appl. Therm. Eng. **30**, 631–638 (2010)
91. J. Shon, H. Kim, K. Lee, Improved heat storage rate for an automobile coolant waste heat recovery system using phase-change material in a fin–tube heat exchanger. Appl. Energy **113**, 680–689 (2014)
92. F. Baldi, C. Gabrielii, F. Melino, M. Bianchi, A preliminary study on the application of thermal storage to merchant ships. Energy Procedia. **75**, 2169–2174 (2015)
93. D.G. Beshore, F.A. Jaeger, E.M. Gartner, Thermal energy storage/waste heat recovery application in the cement industry, in: *Proceedings of First Industrial Energy Technology Conference*, pp. 747–756 (1979)
94. R. De Boer, S.F. Smeding, P.W. Bach, G. De Joode, Heat storage systems for use in an industrial batch process (Results of) a case study, in: Contribution to The Tenth International Conference on Thermal Energy Storage (ECOSTOCK, 2006)
95. M. Barati, S. Esfahani, T.A. Utigard, Energy recovery from high temperature slags. Energy **36**, 5440–5449 (2011)
96. T. Arunkumar, Y. Ao, Z. Luo, L. Zhang, J. Li, D. Denkenberger, J. Wang, Energy efficient materials for solar water distillation-A review. Renew. Sustain. Energy Rev. **115**, 109409 (2019)

97. N.A.S. Elminshawy, F.R. Siddiqui, G.I. Sultan, Development of a desalination system driven by solar energy and low grade waste heat. Energy Convers. Manag. **103**, 28–35 (2015)
98. V.G. Gude, Energy storage for desalination processes powered by renewable energy and waste heat sources. Appl. Energy. **137**, 877–898 (2015)
99. M. Faegh, M.B. Shafii, Experimental investigation of a solar still equipped with an external heat storage system using phase change materials and heat pipes. Desalination **409**, 128–135 (2017)
100. S.W. Sharshir, G. Peng, L. Wu, F.A. Essa, A.E. Kabeel, N. Yang, The effects of flake graphite nanoparticles, phase change material, and film cooling on the solar still performance. Appl. Energy **191**, 358–366 (2017)
101. T. Arunkumar, A.E. Kabeel, Effect of phase change material on concentric circular tubular solar still-Integration meets enhancement. Desalination **414**, 46–50 (2017)
102. G. Wang, Y. Fu, A. Guo, T. Mei, J. Wang, J. Li, X. Wang, Reduced graphene oxide–polyurethane nanocomposite foam as a reusable photoreceiver for efficient solar steam generation. Chem. Mater. **29**, 5629–5635 (2017)
103. G. Ni, N. Miljkovic, H. Ghasemi, X. Huang, S.V. Boriskina, C.-T. Lin, J. Wang, Y. Xu, M.M. Rahman, T. Zhang, Volumetric solar heating of nanofluids for direct vapor generation. Nano Energy **17**, 290–301 (2015)
104. O. Neumann, A.S. Urban, J. Day, S. Lal, P. Nordlander, N.J. Halas, Solar vapor generation enabled by nanoparticles. ACS Nano **7**, 42–49 (2012)
105. Y. Liu, S. Yu, R. Feng, A. Bernard, Y. Liu, Y. Zhang, H. Duan, W. Shang, P. Tao, C. Song, A bioinspired, reusable, paper-based system for high-performance large-scale evaporation. Adv. Mater. **27**, 2768–2774 (2015)
106. Y. Ito, Y. Tanabe, J. Han, T. Fujita, K. Tanigaki, M. Chen, Multifunctional porous graphene for high-efficiency steam generation by heat localization. Adv. Mater. **27**, 4302–4307 (2015)
107. H. Ghasemi, G. Ni, A.M. Marconnet, J. Loomis, S. Yerci, N. Miljkovic, G. Chen, Solar steam generation by heat localization. Nat. Commun. **5**, ncomms5449 (2014)
108. J.J. Yew, S. Rathi, S. Krueger, L.G. Bolton, J.M. Ananny, Connecting multiple accessories to a portable computing device (2011)
109. A. Van Heugten, Systems, devices, and methods for managing camera focus (2017)
110. Z. Ling, Z. Zhang, G. Shi, X. Fang, L. Wang, X. Gao, Y. Fang, T. Xu, S. Wang, X. Liu, Review on thermal management systems using phase change materials for electronic components, Li-ion batteries and photovoltaic modules. Renew. Sustain. Energy Rev. **31**, 427–438 (2014)
111. Y. Deng, C. Feng, E. Jiaqiang, H. Zhu, J. Chen, M. Wen, H. Yin, Effects of different coolants and cooling strategies on the cooling performance of the power lithium ion battery system: a review. Appl. Therm. Eng. **142**, 10–29 (2018)
112. D. Ouyang, M. Chen, Q. Huang, J. Weng, Z. Wang, J. Wang, A review on the thermal hazards of the lithium-ion battery and the corresponding countermeasures. Appl. Sci. **9**, 2483 (2019)
113. P. Huang, A. Verma, D.J. Robles, Q. Wang, P. Mukherjee, J. Sun, Probing the cooling effectiveness of phase change materials on lithium-ion battery thermal response under overcharge condition. Appl. Therm. Eng. **132**, 521–530 (2018)
114. W.G. Alshaer, S.A. Nada, M.A. Rady, E.P. Del Barrio, A. Sommier, Thermal management of electronic devices using carbon foam and PCM/nano-composite. Int. J. Therm. Sci. **89**, 79–86 (2015)
115. A. Arshad, H.M. Ali, M. Ali, S. Manzoor, Thermal performance of phase change material (PCM) based pin-finned heat sinks for electronics devices: Effect of pin thickness and PCM volume fraction. Appl. Therm. Eng. **112**, 143–155 (2017)
116. Z. Luo, H. Cho, X. Luo, K. Cho, System thermal analysis for mobile phone. Appl. Therm. Eng. **28**, 1889–1895 (2008)
117. Z.A. Qureshi, H.M. Ali, S. Khushnood, Recent advances on thermal conductivity enhancement of phase change materials for energy storage system: a review. Int. J. Heat Mass Transf. **127**, 838–856 (2018)
118. H.M. Ali, A. Saieed, W. Pao, M. Ali, Copper foam/PCMs based heat sinks: an experimental study for electronic cooling systems. Int. J. Heat Mass Transf. **127**, 381–393 (2018)

119. B. Agostini, M. Fabbri, J.E. Park, L. Wojtan, J.R. Thome, B. Michel, State of the art of high heat flux cooling technologies. Heat Transf. Eng. **28**, 258–281 (2007)
120. M. Shtein, R. Nadiv, M. Buzaglo, O. Regev, Graphene-based hybrid composites for efficient thermal management of electronic devices. ACS Appl. Mater. Interfaces **7**, 23725–23730 (2015)
121. J. Krishna, P.S. Kishore, A.B. Solomon, Heat pipe with nano enhanced-PCM for electronic cooling application. Exp. Therm. Fluid Sci. **81**, 84–92 (2017)
122. M. Itani, N. Ghaddar, K. Ghali, D. Ouahrani, W. Chakroun, Cooling vest with optimized PCM arrangement targeting torso sensitive areas that trigger comfort when cooled for improving human comfort in hot conditions. Energy Build. **139**, 417–425 (2017)
123. Y. Yusufoglu, T. Apaydin, S. Yilmaz, H.O. Paksoy, Improving performance of household refrigerators by incorporating phase change materials. Int. J. Refrig. **57**, 173–185 (2015)
124. S. Bakhshipour, M.S. Valipour, Y. Pahamli, Parametric analysis of domestic refrigerators using PCM heat exchanger. Int. J. Refrig. **83**, 1–13 (2017)
125. N. Chaiyat, Energy and economic analysis of a building air-conditioner with a phase change material (PCM). Energy Convers. Manag. **94**, 150–158 (2015)
126. M.A. Said, H. Hassan, Effect of using nanoparticles on the performance of thermal energy storage of phase change material coupled with air-conditioning unit. Energy Convers. Manag. **171**, 903–916 (2018)
127. M. Alimohammadi, Y. Aghli, E.S. Alavi, M. Sardarabadi, M. Passandideh-Fard, Experimental investigation of the effects of using nano/phase change materials (NPCM) as coolant of electronic chipsets, under free and forced convection. Appl. Therm. Eng. **111**, 271–279 (2017)
128. B. Praveen, S. Suresh, Experimental study on heat transfer performance of neopentyl glycol/CuO composite solid-solid PCM in TES based heat sink. Eng. Sci. Technol. An Int. J. **21**, 1086–1094 (2018)
129. Z. Huang, N. Xie, X. Zheng, X. Gao, X. Fang, Y. Fang, Z. Zhang, Experimental and numerical study on thermal performance of Wood's alloy/expanded graphite composite phase change material for temperature control of electronic devices. Int. J. Therm. Sci. **135**, 375–385 (2019)
130. A. Farzanehnia, M. Khatibi, M. Sardarabadi, M. Passandideh-Fard, Experimental investigation of multiwall carbon nanotube/paraffin based heat sink for electronic device thermal management. Energy Convers. Manag. **179**, 314–325 (2019)
131. B. Praveen, S. Suresh, V. Pethurajan, Heat transfer performance of graphene nano-platelets laden micro-encapsulated PCM with polymer shell for thermal energy storage based heat sink. Appl. Therm. Eng. **156**, 237–249 (2019)
132. P.M. Congedo, M. Malvoni, M. Mele, M.G. De Giorgi, Performance measurements of monocrystalline silicon PV modules in South-eastern Italy. Energy Convers. Manag. **68**, 1–10 (2013). https://doi.org/10.1016/j.enconman.2012.12.017
133. M. Chandrasekar, S. Suresh, T. Senthilkumar, M. Ganesh Karthikeyan, Passive cooling of standalone flat PV module with cotton wick structures. Energy Convers. Manag. **71**, 43–50 (2013). https://doi.org/10.1016/j.enconman.2013.03.012
134. a Q. Malik, M. Fauzi, Performance of single crystal silicon photovoltaic module in Bruneian Climate, pp. 179–188 (2010)
135. H. Jiang, L. Lu, K. Sun, Experimental investigation of the impact of airborne dust deposition on the performance of solar photovoltaic (PV) modules. Atmos. Environ. **45**, 4299–4304 (2011). https://doi.org/10.1016/j.atmosenv.2011.04.084
136. D.M. Chapin, C.S. Fuller, G.L. Pearson, A new silicon p-n junction photocell for converting solar radiation into electrical power. J. Appl. Phys. **25**, 676–677 (1954)
137. S.R. Kurtz, Opportunities and challenges for development of a mature concentrating photovoltaic power industry (2009)
138. A. Luque, S. Hegedus, *Handbook of photovoltaic science and engineering* (Wiley, 2011)
139. P. Huen, W.A. Daoud, Advances in hybrid solar photovoltaic and thermoelectric generators. Renew. Sustain. Energy Rev. **72**, 1295–1302 (2017)
140. P.G.V. Sampaio, M.O.A. González, R.M. de Vasconcelos, M.A.T. dos Santos, J.C. de Toledo, J.P.P. Pereira, Photovoltaic technologies: mapping from patent analysis. Renew. Sustain. Energy Rev. **93**, 215–224 (2018)

141. K. Agroui, A.H. Arab, M. Pellegrino, Indoor and outdoor photovoltaic modules Performances based on thin films solar cells **14**, 469–480 (2011)
142. J.Y. Ye, K. Ding, T. Reindl, A.G. Aberle, Outdoor PV module performance under fluctuating irradiance conditions in tropical climates. Energy Procedia **33**, 238–247 (2013). https://doi.org/10.1016/j.egypro.2013.05.064
143. L.S. Pantić, T.M. Pavlović, D.D. Milosavuević, A practical field study of performances of solar modules at various positions in Serbia. Therm. Sci. **19**, S511–S523 (2015). https://doi.org/10.2298/TSCI140313081P
144. M.E. Başoılu, A. Kazdaloılu, T. Erfidan, M.Z. Bilgin, B. Cąkir, Performance analyzes of different photovoltaic module technologies under¯zmit, Kocaeli climatic conditions. Renew. Sustain. Energy Rev. **52**, 357–365 (2015). https://doi.org/10.1016/j.rser.2015.07.108
145. R. Eke, H. Demircan, Performance analysis of a multi crystalline Si photovoltaic module under Mugla climatic conditions in Turkey. Energy Convers. Manag. **65**, 580–586 (2013). https://doi.org/10.1016/j.enconman.2012.09.007
146. C. Cañete, J. Carretero, M. Sidrach-de-Cardona, Energy performance of different photovoltaic module technologies under outdoor conditions. Energy **65**, 295–302 (2014). https://doi.org/10.1016/j.energy.2013.12.013
147. M.R. Abdelkader, F. Sharaf, A comparative Analysis of the Performance of Monocrystalline and Multiycrystalline PV Cells in Semi Arid Climate Conditions : the Case of Jordan. Jordan J. Mech. Ind. Eng. **4**, 543–552 (2010)
148. N. Amin, C.W. Lung, K. Sopian, A practical field study of various solar cells on their performance in Malaysia. Renew. Energy **34**, 1939–1946 (2009). https://doi.org/10.1016/j.renene.2008.12.005
149. O.M. Midtgard, T.O. Sætre, G. Yordanov, A.G. Imenes, C.L. Nge, A qualitative examination of performance and energy yield of photovoltaic modules in southern Norway. Renew. Energy **35**, 1266–1274 (2010). https://doi.org/10.1016/j.renene.2009.12.002
150. S. Dubey, J.N. Sarvaiya, B. Seshadri, Temperature dependent photovoltaic (PV) efficiency and its effect on PV production in the world—a review. Energy Procedia **33**, 311–321 (2013). https://doi.org/10.1016/j.egypro.2013.05.072
151. E. Skoplaki, J.A. Palyvos, On the temperature dependence of photovoltaic module electrical performance: a review of efficiency/power correlations. Sol. Energy **83**, 614–624 (2009). https://doi.org/10.1016/j.solener.2008.10.008
152. N. Suwapaet, P. Boonla, The investigation of produced power output during high operating temperature occurrences of monocrystalline and amorphous photovoltaic modules. Energy Procedia **52**, 459–465 (2014). https://doi.org/10.1016/j.egypro.2014.07.098
153. P. Singh, S.N. Singh, M. Lal, M. Husain, Temperature dependence of I-V characteristics and performance parameters of silicon solar cell. Sol. Energy Mater. Sol. Cells. **92**, 1611–1616 (2008). https://doi.org/10.1016/j.solmat.2008.07.010
154. S. Mekhilef, R. Saidur, M. Kamalisarvestani, Effect of dust, humidity and air velocity on efficiency of photovoltaic cells. Renew. Sustain. Energy Rev. **16**, 2920–2925 (2012). https://doi.org/10.1016/j.rser.2012.02.012
155. B. Zalba, J.M. Marın, L.F. Cabeza, H. Mehling, Review on thermal energy storage with phase change: materials, heat transfer analysis and applications. Appl. Therm. Eng. **23**, 251–283 (2003)
156. M. Mani, R. Pillai, Impact of dust on solar photovoltaic (PV) performance: Research status, challenges and recommendations. Renew. Sustain. Energy Rev. **14**, 3124–3131 (2010). https://doi.org/10.1016/j.rser.2010.07.065
157. M. Fuentes, G. Nofuentes, J. Aguilera, D.L. Talavera, M. Castro, Application and validation of algebraic methods to predict the behaviour of crystalline silicon PV modules in Mediterranean climates. Sol. Energy **81**, 1396–1408 (2007). https://doi.org/10.1016/j.solener.2006.12.008
158. M. Torres-Ramírez, G. Nofuentes, J.P. Silva, S. Silvestre, J.V. Muñoz, Study on analytical modelling approaches to the performance of thin film PV modules in sunny inland climates. Energy **73**, 731–740 (2014). https://doi.org/10.1016/j.energy.2014.06.077

159. G. Nofuentes, M. Fuentes, J. Aguilera, J.V. Muñoz, An assessment on simple modeling approaches to the electric behavior of two cis pv modules in a sunny climate. J. Sol. Energy Eng. Trans. ASME **131**, 0310131–03101310 (2009). https://doi.org/10.1115/1.3142800
160. U.K. Mirza, M. Mercedes Maroto-Valer, N. Ahmad, Status and outlook of solar energy use in Pakistan. Renew. Sustain. Energy Rev. **7**, 501–514 (2003). https://doi.org/10.1016/j.rser.2003.06.002
161. I. Ulfat, F. Javed, F.A. Abbasi, F. Kanwal, A. Usman, M. Jahangir, F. Ahmed, Estimation of solar energy potential for Islamabad, Pakistan. Energy Procedia **18**, 1496–1500 (2012). https://doi.org/10.1016/j.egypro.2012.05.166
162. D.B. Richardson, L.D.D. Harvey, Strategies for correlating solar PV array production with electricity demand. Renew. Energy **76**, 432–440 (2015). https://doi.org/10.1016/j.renene.2014.11.053
163. B. Parida, S. Iniyan, R. Goic, A review of solar photovoltaic technologies. Renew. Sustain. Energy Rev. **15**, 1625–1636 (2011). https://doi.org/10.1016/j.rser.2010.11.032
164. H.G. Teo, P.S. Lee, M.N.A. Hawlader, An active cooling system for photovoltaic modules. Appl. Energy **90**, 309–315 (2012)
165. D.J. Yang, Z.F. Yuan, P.H. Lee, H.M. Yin, Simulation and experimental validation of heat transfer in a novel hybrid solar panel. Int. J. Heat Mass Transf. **55**, 1076–1082 (2012). https://doi.org/10.1016/j.ijheatmasstransfer.2011.10.003
166. A. Makki, S. Omer, H. Sabir, Advancements in hybrid photovoltaic systems for enhanced solar cells performance. Renew. Sustain. Energy Rev. **41**, 658–684 (2015). https://doi.org/10.1016/j.rser.2014.08.069
167. C. Schwingshackl, M. Petitta, J.E. Wagner, G. Belluardo, D. Moser, M. Castelli, M. Zebisch, A. Tetzlaff, Wind effect on PV module temperature: analysis of different techniques for an accurate estimation. Energy Procedia **40**, 77–86 (2013). https://doi.org/10.1016/j.egypro.2013.08.010
168. E. Kermani, Manifold micro-channel cooling of photovoltaic cells for high efficiency solar energy conversion systems (2008)
169. M.M. Rahman, M. Hasanuzzaman, N.A. Rahim, Effects of various parameters on PV-module power and efficiency. Energy Convers. Manag. **103**, 348–358 (2015)
170. O. Dupré, R. Vaillon, M.A. Green, Physics of the temperature coefficients of solar cells. Sol. Energy Mater. Sol. Cells. **140**, 92–100 (2015)
171. J.W. Stultz, L.C. Wen, Thermal performance testing and analysis of photovoltaic modules in natural sunlight. LSA Task Rep. **5101**, 31 (1977)
172. A. Hasan, S.J. McCormack, M.J. Huang, B. Norton, Evaluation of phase change materials for thermal regulation enhancement of building integrated photovoltaics. Sol. Energy **84**, 1601–1612 (2010)
173. C.J. Smith, P.M. Forster, R. Crook, Global analysis of photovoltaic energy output enhanced by phase change material cooling. Appl. Energy **126**, 21–28 (2014)
174. U. Stritih, Increasing the efficiency of PV panel with the use of PCM. Renew. Energy **97**, 671–679 (2016)
175. S. Maiti, S. Banerjee, K. Vyas, P. Patel, P.K. Ghosh, Self regulation of photovoltaic module temperature in V-trough using a metal–wax composite phase change matrix. Sol. Energy **85**, 1805–1816 (2011)
176. P.H. Biwole, P. Eclache, F. Kuznik, Phase-change materials to improve solar panel's performance. Energy Build. **62**, 59–67 (2013)
177. M.J. Huang, P.C. Eames, B. Norton, N.J. Hewitt, Natural convection in an internally finned phase change material heat sink for the thermal management of photovoltaics. Sol. Energy Mater. Sol. Cells. **95**, 1598–1603 (2011)
178. M.C. Browne, K. Lawlor, A. Kelly, B. Norton, S.J. Mc Cormack, Indoor characterisation of a photovoltaic/thermal phase change material system. Energy Procedia. **70**, 163–171 (2015)
179. M.C. Browne, B. Norton, S.J. McCormack, Phase change materials for photovoltaic thermal management. Renew. Sustain. Energy Rev. **47**, 762–782 (2015)

180. P. Atkin, M.M. Farid, Improving the efficiency of photovoltaic cells using PCM infused graphite and aluminium fins. Sol. Energy. **114**, 217–228 (2015)
181. A.H.A. Al-Waeli, K. Sopian, M.T. Chaichan, H.A. Kazem, A. Ibrahim, S. Mat, M.H. Ruslan, Evaluation of the nanofluid and nano-PCM based photovoltaic thermal (PVT) system: an experimental study. Energy Convers. Manag. **151**, 693–708 (2017)
182. M.K.A. Ali, H. Xianjun, L. Mai, C. Bicheng, R.F. Turkson, C. Qingping, Reducing frictional power losses and improving the scuffing resistance in automotive engines using hybrid nanomaterials as nano-lubricant additives. Wear **364**, 270–281 (2016)
183. A.H.A. Al-Waeli, H.A. Kazem, M.T. Chaichan, K. Sopian, Experimental investigation of using nano-PCM/nanofluid on a photovoltaic thermal system (PVT): technical and economic study. Therm. Sci. Eng. Prog. **11**, 213–230 (2019)
184. N.S. Dhaidan, J.M. Khodadadi, T.A. Al-Hattab, S.M. Al-Mashat, Experimental and numerical investigation of melting of NePCM inside an annular container under a constant heat flux including the effect of eccentricity. Int. J. Heat Mass Transf. **67**, 455–468 (2013)
185. D. Groulx, Numerical study of nano-enhanced PCMs: are they worth it? in *Proceedings of the 1st Thermal and Fluid Engineering Summer Conference (TFESC)*, New York, Aug 2015, pp. 9–12
186. R. Parameshwaran, K. Deepak, R. Saravanan, S. Kalaiselvam, Preparation, thermal and rheological properties of hybrid nanocomposite phase change material for thermal energy storage. Appl. Energy **115**, 320–330 (2014)
187. M. Amin, N. Putra, E.A. Kosasih, E. Prawiro, R.A. Luanto, T.M.I. Mahlia, Thermal properties of beeswax/graphene phase change material as energy storage for building applications. Appl. Therm. Eng. **112**, 273–280 (2017)
188. Z. Luo, H. Zhang, X. Gao, T. Xu, Y. Fang, Z. Zhang, Fabrication and characterization of form-stable capric-palmitic-stearic acid ternary eutectic mixture/nano-SiO2 composite phase change material. Energy Build. **147**, 41–46 (2017)
189. S. Sharma, L. Micheli, W. Chang, A.A. Tahir, K.S. Reddy, T.K. Mallick, Nano-enhanced phase change material for thermal management of BICPV. Appl. Energy. **208**, 719–733 (2017)
190. Z. Ma, W. Lin, M.I. Sohel, Nano-enhanced phase change materials for improved building performance. Renew. Sustain. Energy Rev. **58**, 1256–1268 (2016)
191. S.C. Lin, H.H. Al-Kayiem, Evaluation of copper nanoparticles—Paraffin wax compositions for solar thermal energy storage. Sol. Energy **132**, 267–278 (2016). https://doi.org/10.1016/j.solener.2016.03.004
192. N. Abdollahi, M. Rahimi, Potential of water natural circulation coupled with nano-enhanced PCM for PV module cooling. Renew. Energy **147**, 302–309 (2020)
193. H.E. Abdelrahman, M.H. Wahba, H.A. Refaey, M. Moawad, N.S. Berbish, Performance enhancement of photovoltaic cells by changing configuration and using PCM (RT35HC) with nanoparticles Al_2O_3. Sol. Energy **177**, 665–671 (2019)
194. M. Al-Jethelah, S.H. Tasnim, S. Mahmud, A. Dutta, Nano-PCM filled energy storage system for solar-thermal applications. Renew. Energy **126**, 137–155 (2018)
195. S.Y. Wu, H. Wang, S. Xiao, D.S. Zhu, An investigation of melting/freezing characteristics of nanoparticle-enhanced phase change materials. J. Therm. Anal. Calorim. **110**, 1127–1131 (2011)
196. T.-P. Teng, C.-C. Yu, Characteristics of phase-change materials containing oxide nano-additives for thermal storage. Nanoscale Res. Lett. **7**, 611 (2012)
197. T. Oya, T. Nomura, M. Tsubota, N. Okinaka, T. Akiyama, Thermal conductivity enhancement of erythritol as PCM by using graphite and nickel particles. Appl. Therm. Eng. **61**, 825–828 (2013)
198. Q. He, S. Wang, M. Tong, Y. Liu, Experimental study on thermophysical properties of nanofluids as phase-change material (PCM) in low temperature cool storage. Energy Convers. Manag. **64**, 199–205 (2012)
199. Y. Zeng, L.-W. Fan, Y.-Q. Xiao, Z.-T. Yu, K.-F. Cen, An experimental investigation of melting of nanoparticle-enhanced phase change materials (NePCMs) in a bottom-heated vertical cylindrical cavity. Int. J. Heat Mass Transf. **66**, 111–117 (2013)

200. Y. Cui, C. Liu, S. Hu, X. Yu, The experimental exploration of carbon nanofiber and carbon nanotube additives on thermal behavior of phase change materials. Sol. Energy Mater. Sol. Cells. **95**, 1208–1212 (2011)
201. M. Li, A nano-graphite/paraffin phase change material with high thermal conductivity. Appl. Energy **106**, 25–30 (2013)
202. A.H. Ali, S.I. Ibrahim, Q.A. Jawad, R.S. Jawad, M.T. Chaichan, Effect of nanomaterial addition on the thermos physical properties of Iraqi paraffin wax. Case Stud. Therm. Eng. 100537 (2019)
203. M. Kazemi, A. Kianifar, H. Niazmand, Nanoparticle loading effect on the performance of the paraffin thermal energy storage material for building applications. J. Therm. Anal. Calorim. 1–7 (n.d.)